METHODS IN MOLECULAR BIOLOGY

Series Editor
John M. Walker
School of Life Sciences
University of Hertfordshire
Hatfield, Hertfordshire, AL10 9AB, UK

For further volumes:
http://www.springer.com/series/7651

Biomimetics and Stem Cells

Methods and Protocols

Edited by

Gordana Vunjak-Novakovic

Department of Biomedical Engineering and Department of Medicine, Columbia University, New York, NY, USA

Kursad Turksen

Sprott Centre for Stem Cell Research, Ottawa Hospital Research Institute, Ottawa, ON, Canada

 Humana Press

Editors
Gordana Vunjak-Novakovic
Department of Biomedical Engineering
  and Department of Medicine
Columbia University
New York, NY, USA

Kursad Turksen
Ottawa Hospital Research Institute
Sprott Centre for Stem Cell Research
Regenerative Medicine Program
Ottawa, ON, Canada

ISSN 1064-3745   ISSN 1940-6029 (electronic)
ISBN 978-1-4939-1331-2   ISBN 978-1-4939-1332-9 (eBook)
DOI 10.1007/978-1-4939-1332-9
Springer New York Heidelberg Dordrecht London

Library of Congress Control Number: 2014942165

Printed on acid-free paper

Humana Press is a brand of Springer
Springer is part of Springer Science+Business Media (www.springer.com)

Preface

The term *biomimetics* comes from the Greek words *bios*, meaning life, and *mimesis*, meaning imitation. Leonardo da Vinci was probably the first one to apply this principle in his attempts to make a flying machine by studying the birds. The term was coined in 1950s by Otto Schmitt when he attempted to engineer an electronic device that replicated signal propagation in a squid nerve. Today, biomimetics is a broad concept viewed as the use of synthetic systems that mimic biological materials, mechanisms, and processes.

We focus here on the use of biomimetic systems for stem cells, for both scientific and practical reasons. It is well known that living cells respond to their environment, in vitro and in vivo, in a way that determines their fate and function. Stem cells are more responsive to the external stimuli than other cell types, due to their "plasticity," which is in turn the basis of their use for tissue regeneration and developmental studies. Recent advances in tissue engineering are enabling us to "instruct" the stem cells towards differentiating into the right phenotypes, in the right place and at the right time. Moving toward meeting this challenge, biomimetic environments are being designed to recapitulate in vitro the combinations of factors known to guide tissue development and regeneration in vivo and thereby help unlock the full potential of the stem cells. We have selected a series of approaches demonstrating the role and value of biomimetics for the better understanding of stem cell behavior and the acceleration of their application in regenerative medicine.

We sincerely thank all authors for their outstanding contributions. We also thank Dr. John Walker, the Editor in Chief of the MIMB series, for his continued support. We are grateful to Patrick Martin, Editor of the Springer Protocols series, and to David Casey for ensuring that this volume will meet the highest standards.

New York, NY, USA   *Gordana Vunjak-Novakovic*
Ottawa, ON, Canada   *Kursad Turksen*

Contents

Contributors

AHMAD E. ABU-HAKMEH • *Department of Biomedical Engineering, Center for Biotechnology & Interdisciplinary Studies, Rensselaer Polytechnic Institute, Troy, NY, USA*

J. STEWART AITCHISON • *The Edward S. Rogers Sr. Department of Electrical and Computer Engineering, University of Toronto, Toronto, ON, Canada*

LUIS F. ALONZO • *Department of Biomedical Engineering, University of California, Irvine, CA, USA*

STEPHEN F. BADYLAK • *Department of Surgery, McGowan Institute for Regenerative Medicine, University of Pittsburgh, Pittsburgh, PA, USA; Department of Bioengineering, McGowan Institute for Regenerative Medicine, University of Pittsburgh, Pittsburgh, PA, USA*

SARINDR BHUMIRATANA • *Department of Biomedical Engineering, Columbia University, New York, NY, USA*

DANIELLE R. BOGDANOWICZ • *Department of Biomedical Engineering, Columbia University, New York, NY, USA*

ANTHONY CONWAY • *Department of Chemical and Biomolecular Engineering, University of California, Berkeley, CA, USA; Department of Bioengineering, University of California, Berkeley, CA, USA*

STEVEN C. GEORGE • *Department of Biomedical Engineering, The Edwards Life Sciences Center for Advanced Cardiovascular Technology, University of California, Irvine, CA, USA; Department of Chemical Engineering and Materials Science, University of California, Irvine, CA, USA; Department of Medicine, The Edwards Life Sciences Center for Advanced Cardiovascular Technology, University of California, Irvine, CA, USA*

SHARON GERECHT • *Department of Chemical and Biomolecular Engineering, Johns Hopkins Physical Sciences, and Oncology Center and Institute for NanoBioTechnology, Johns Hopkins University, Baltimore, MD, USA; Department of Materials Science and Engineering, Johns Hopkins University, Baltimore, MD, USA*

WARREN L. GRAYSON • *Department of Biomedical Engineering, Translational Tissue Engineering Center, Johns Hopkins University, Baltimore, MD, USA*

MUSTAFA O. GULER • *UNAM-Institute of Materials Science and Nanotechnology, Bilkent University, Bilkent, Ankara, Turkey*

JEFFREY O. HOLLINGER • *Department of Biomedical Engineering, Carnegie Mellon University, Pittsburgh, PA, USA; Department of Biological Sciences, Carnegie Mellon University, Pittsburgh, PA, USA*

SRAVANTI KUSUMA • *Department of Chemical and Biomolecular Engineering, Johns Hopkins Physical Sciences, and Oncology Center and Institute for NanoBioTechnology, Johns Hopkins University, Baltimore, MD, USA*

KANGAE LEE • *Department of Chemical & Biological Engineering, School of Engineering and Applied Science, Princeton University, Princeton, NJ, USA*

CORNELIA LEE-THEDIECK • *Institute of Functional Interfaces, Karlsruhe Institute of Technology (KIT), Eggenstein-Leopoldshafen, Germany*

CAMILA LONDONO • *Institute of Biomaterials and Biomedical Engineering, University of Toronto, Toronto, ON, Canada*

RICARDO LONDONO • *Department of Cellular and Molecular Pathology, McGowan Institute for Regenerative Medicine, University of Pittsburgh, Pittsburgh, PA, USA*

HELEN H. LU • *Department of Biomedical Engineering, Columbia University, New York, NY, USA*

ZUFU LU • *Biomaterials and Tissue Engineering Research Unit, School of AMME, The University of Sydney, Sydney, Australia*

PETRA B. LÜCKER • *Department of Chemical Engineering and Applied Chemistry, Institute of Biomaterials and Biomedical Engineering, University of Toronto, Toronto, ON, Canada*

BRIA MACKLIN • *Department of Chemical and Biomolecular Engineering, Johns Hopkins Physical Sciences, and Oncology Center and Institute for NanoBioTechnology, Johns Hopkins University, Baltimore, MD, USA*

BUSRA MAMMADOV • *UNAM-Institute of Materials Science and Nanotechnology, Bilkent University, Bilkent, Ankara, Turkey*

DARJA MAROLT • *The New York Stem Cell Foundation Research Institute, New York, NY, USA*

ANA M. MARTINS • *3B's Research Group—Biomaterials, Biodegradables and Biomimetics, University of Minho, Guimarães, Portugal*

ALISON P. MCGUIGAN • *Department of Chemical Engineering and Applied Chemistry, Institute of Biomaterials and Biomedical Engineering, University of Toronto, Toronto, ON, Canada*

MONICA L. MOYA • *Department of Biomedical Engineering, University of California, Irvine, CA, USA*

CELESTE M. NELSON • *Department of Chemical & Biological Engineering, School of Engineering and Applied Science, Princeton University, Princeton, NJ, USA*

GIUSEPPE MARIA DEPEPPO • *The New York Stem Cell Foundation Research Institute, New York, NY, USA*

ANNAMARIJA RAIC • *RWTH Aachen University, Aachen, Germany*

RUI L. REIS • *3B's Research Group—Biomaterials, Biodegradables and Biomimetics, University of Minho, Guimarães, Portugal*

LISA RÖDLING • *Institute of Functional Interfaces, Karlsruhe Institute of Technology (KIT), Eggenstein-Leopoldshafen, Germany*

SEYED-IMAN ROOHANI-ESFAHANI • *Biomaterials and Tissue Engineering Research Unit, School of AMME, The University of Sydney, Sydney, Australia*

SUTHAMATHY SATHANANTHAN • *The Edward S. Rogers Sr. Department of Electrical and Computer Engineering, University of Toronto, Toronto, ON, Canada*

DAVID V. SCHAFFER • *Department of Chemical and Biomolecular Engineering, University of California, Berkeley, CA, USA; Department of Bioengineering, University of California, Berkeley, CA, USA*

ARUN R. SHRIVATS • *Department of Biomedical Engineering, Carnegie Mellon University, Pittsburgh, PA, USA; Department of Biological Sciences, Carnegie Mellon University, Pittsburgh, PA, USA*

BRIAN M. SICARI • *Department of Cellular and Molecular Pathology, McGowan Institute for Regenerative Medicine, University of Pittsburgh, Pittsburgh, PA, USA*

JOHN SOLEAS • *Institute of Biomaterials and Biomedical Engineering, University of Toronto, Toronto, ON, Canada*

DAWN P. SPELKE • *Department of Chemical and Biomolecular Engineering, University of California, Berkeley, CA, USA; Department of Bioengineering, University of California, Berkeley, CA, USA*

AYSE B. TEKINAY • *UNAM-Institute of Materials Science and Nanotechnology, Bilkent University, Bilkent, Ankara, Turkey*

JOSHUA P. TEMPLE • *Department of Biomedical Engineering, Translational Tissue Engineering Center, Johns Hopkins University, Baltimore, MD, USA*

GORDANA VUNJAK-NOVAKOVIC • *Department of Biomedical Engineering and Department of Medicine, Columbia University, New York, NY, USA*

LEO Q. WAN • *Department of Biomedical Engineering, Center for Biotechnology & Interdisciplinary Studies, Rensselaer Polytechnic Institute, Troy, NY, USA*

KEITH YEAGER • *Department of Biomedical Engineering, Columbia University, New York, NY, USA*

LI ZHANG • *McGowan Institute for Regenerative Medicine, University of Pittsburgh, Pittsburgh, PA, USA*

HALA ZREIQAT • *Biomaterials and Tissue Engineering Research Unit, School of AMME, The University of Sydney, Sydney, Australia*

Methods in Molecular Biology (2014) 1202: 1–9
DOI 10.1007/7651_2013_39
© Springer Science+Business Media New York 2013
Published online: 24 October 2013

Derivation and Network Formation of Vascular Cells from Human Pluripotent Stem Cells

Sravanti Kusuma, Bria Macklin, and Sharon Gerecht

Abstract

As the lifeline of almost all living tissues, blood vessels are a major focus of tissue-regenerative therapies. Rebuilding blood vessels has vast implications for the study of vascular growth and treatment of diseases in which vascular function is compromised. Toward this end, human pluripotent stem cells have been widely studied for their differentiation capacity toward vascular lineages. We demonstrate methods to derive a bicellular population of early specialized vascular cells from human pluripotent stem cells, to differentiate these toward mature endothelial cells and pericytes, and to utilize a collagen scaffold to facilitate organization into vascular networks.

Keywords: Endothelial cells, Hydrogels, Pericytes, Pluripotent stem cells

1 Introduction

Recreating functional blood vessels is desirable for the study of three-dimensional vascular biology toward the goals of regenerating tissue and treating diseases in which vascular function is compromised (1). Cell-based therapies are gaining momentum as permanent and effective strategies to rebuild vasculature; however, harvested somatic cells are difficult to isolate and expand in vitro, retain the potential for disease or damage, and may damage the donor site. As an alternative, human pluripotent stem cells (hPSCs), which can differentiate into every cell type of the body, have been a widely studied stem cell source for vascular reconstruction (2). Especially with the advent of human induced PSCs (hiPSCs), we now have the capability to derive vascular cells in a patient-specific manner (3).

The vasculature is a multicellular tissue in which no cell type can function alone. In three-dimensional (3D) environments, endothelial cells (ECs), which compose the vasculature's inner lining, form nascent endothelial tubes that regress without pericyte recruitment (4, 5). Biomimetic 3D environments such as polymeric hydrogels have been studied to recreate this process in vitro to elucidate mechanisms of vascular growth and to serve as scaffolds for

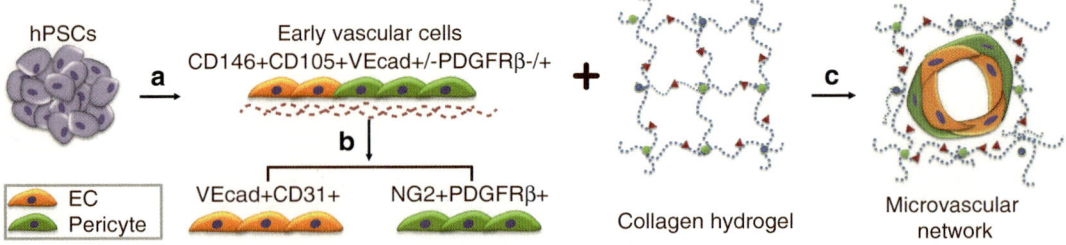

Fig. 1 Schema for self-assembled vascular derivatives. (**a**) hPSCs are differentiated toward early vascular cells (EVCs) that can be (**b**) matured into functional ECs and pericytes. (**c**) Derived EVCs are embedded within a collagen matrix to facilitate organization into vascular networks. Adapted from (10)

implantation (6–8). Because it makes up the majority of stroma, collagen has been a widely studied scaffold to recapitulate physiological environments in which to study vascular cell behavior. These studies have confirmed that the presence of both ECs and a supporting cell type such as pericytes greatly augments vascular growth and durability (4, 9).

Currently, there is a major focus to derive vascular cells that are amenable to clinical translation. In this chapter, we present a clinically relevant protocol to differentiate hPSCs toward early vascular cells (EVCs), a specialized vascular population made up of early ECs and pericytes (Fig. 1a) (10). We also describe the protocols to further differentiate the EVC subpopulations toward mature ECs and pericytes (Fig. 1b). Furthermore, we demonstrate assembly of the bicellular population within a three-dimensional collagen matrix into microvascular networks (Fig. 1c).

2 Materials

Prepare and store all medium at 4 °C. Medium should be warmed to 37 °C prior to use.

1. Culture of undifferentiated hPSCs to ~80 % confluence of colonies, on MEF feeder layers or feeder-free conditions in 6-well culture dishes.

2. PBS.

3. 15 and 50 ml sterile conical tubes.

4. 5 and 10 ml sterile serological pipettes.

5. 10 and 1,000 µl micropipette tips and pipettors.

6. EDTA: 5 mM EDTA in PBS, supplemented with 1 % (v/v) Hyclone FBS and 0.1 % (v/v) β-mercaptoethanol.

7. 40 µm strainer.

8. Hemocytometer.

9. Collagen type IV-coated 6-well plates.

10. Early differentiation medium: Alpha minimum essential medium supplemented with 10 % Hyclone FBS and 0.1 mM β-mercaptoethanol.

11. TrypLE™.

12. EVC differentiation medium: Endothelial basal growth medium supplemented with 2 % serum (PromoCell), 50 ng/ml vascular endothelial growth factor (VEGF), 10 μM TGFβ inhibitor, SB431542, and 0.1 % penicillin–streptomycin antibiotic.

13. MACS buffer: 0.5 % Bovine serum albumin (BSA) and 2 mM EDTA in PBS.

14. VEcad-PE-conjugated antibody.

15. Anti-PE microbeads (Miltenyi Biotec).

16. MACS MS column (Miltenyi Biotec).

17. Pericyte maturation medium: DMEM (without sodium pyruvate) supplemented with 10 % heat-inactivated FBS.

18. Tissue culture-treated 6-well plates.

19. 96-well plates.

20. 1 % Polyethylenimine (PEI) solution: 1 g PEI diluted in 100 ml distilled water.

21. 0.1 % glutaraldehyde solution: 0.1 % (v/v) glutaraldehyde in distilled water.

22. Medium 199 $1\times$.

23. Medium 199 $10\times$.

24. 1 M NaOH.

25. Rat tail type I collagen solution: 7.1 mg/ml collagen in 0.1 % acetic acid.

26. Collagen gel culture medium: Endothelial basal growth medium supplemented with 2 % serum, 50 ng/ml VEGF, stromal derived factor (SDF), stem cell factor (SCF), and interleukin-3 (IL3).

3 Methods

Carry out all procedures at room temperature and in a sterile laminar flow cabinet, unless otherwise specified.

3.1 EVC Derivation

1. Treat undifferentiated hPSC colonies grown in 6-well plates with 0.5 ml EDTA per well (see Note 1).

2. Incubate in 37 °C incubator for 10–15 min.

3. Add 1 ml early differentiation medium per well. Gently scratch the well's surface with a serological pipette to ensure complete cellular detachment.

4. Collect cell suspension in sterile 15 ml conical tube, and centrifuge for 3 min at $240 \times g$ and 4 °C.

5. Aspirate supernatant, and resuspend pellet in early differentiation medium. Repeatedly pipette the cell suspension up and down using a 1,000 μm micropipette tip to obtain a single-cell suspension.

6. Filter the cell suspension through a 40 μm strainer to remove aggregates.

7. Count cells, and calculate the required number of wells to culture cells at a density of 500,000 cells per well.

8. Dilute single-cell suspension such that they are at a final volume of 1 ml per well.

9. Add 1 ml early differentiation medium per well on a collagen type IV-coated 6-well plate.

10. Add 1 ml of cell suspension per well onto collagen IV-coated dishes. Culture in 37 °C incubator at 5 % CO_2. This day of seeding is designated "day 0."

11. Change medium daily.

12. On day 6, remove early differentiation medium and add PBS.

13. Aspirate PBS.

14. Add 0.5 ml TrypLE per well.

15. Incubate in 37 °C incubator. After 5 min, check under light microscope that cells have detached.

16. Add 1 ml early differentiation medium per well, and collect cell suspension in a 15 ml sterile conical tube.

17. Centrifuge for 3 min at $240 \times g$ and 4 °C.

18. Aspirate supernatant, and re-suspend pellet in EVC differentiation medium. Repeatedly pipette the cell suspension up and down using a 1,000 μm micropipette tip to obtain a single-cell suspension.

19. Filter cell suspension through a 40 μm strainer to remove aggregates.

20. Count cells, and calculate the required number of wells to culture cells at a density of 12,500 cells per cm^2.

21. Dilute cell suspension such that they are at a final volume of 1 ml per well.

22. Add 1 ml EVC differentiation medium per well on a collagen IV-coated dish.

23. Plate 1 ml of cell suspension per well onto collagen IV-coated dishes. Culture in 37 °C incubator at 5 % CO_2 (see Note 2).

24. Change medium every other day.

25. On day 12, remove EVC differentiation medium and add PBS.

26. Aspirate PBS.

27. Add TrypLE.

28. Incubate in 37 °C incubator. After 5 min, check under light microscope that cells have detached.

29. Add 1 ml EVC differentiation medium per well, and collect cell suspension in a sterile tube.

30. Centrifuge for 3 min at 240 × g and 4 °C.

31. Aspirate supernatant, and re-suspend pellet in appropriate medium.

32. Strain cell suspension through a 40 μm strainer to remove aggregates.

33. Count cells, and proceed with desired downstream analysis, maturation toward ECs (Section 3.2), or maturation toward pericytes (Section 3.3).

3.2 EC Maturation

1. Re-suspend no more than 1 × 10^6 cells in 100 μl MACS buffer.

2. Add 10 μl of VEcad-PE antibody. Vortex to mix.

3. Incubate on ice and in the dark for 30–45 min.

4. Centrifuge cells for 3 min at 240 × g and 4 °C.

5. Remove supernatant.

6. Resuspend in 1 ml MACS buffer, and centrifuge for 3 min at 240 × g and 4 °C.

7. Remove supernatant.

8. Add 80 μl MACS buffer and 20 μl anti-PE microbeads.

9. Incubate for 15 min in the fridge in the dark.

10. Repeat steps 4–7.

11. Re-suspend in 1 ml MACS buffer.

12. Place MS MACS column in magnetic column holder and a 15 ml conical tube to collect eluate.

13. Add 1 ml MACS buffer to column to wet.

14. After buffer has completely eluted, add cells through a 40 μm strainer.

15. Allow solution to completely pass through column, and then add 0.5 ml MACS buffer to the column three times, adding only when the previous buffer is completely eluted.

16. After three washes, remove column from magnet and place in a sterile 15 ml tube.

17. Add 1 ml MACS buffer, and use plunger to forcefully remove cells in one motion (see Note 3).

18. Centrifuge cells for 3 min and 240 × g and 4 °C.

19. Remove supernatant.

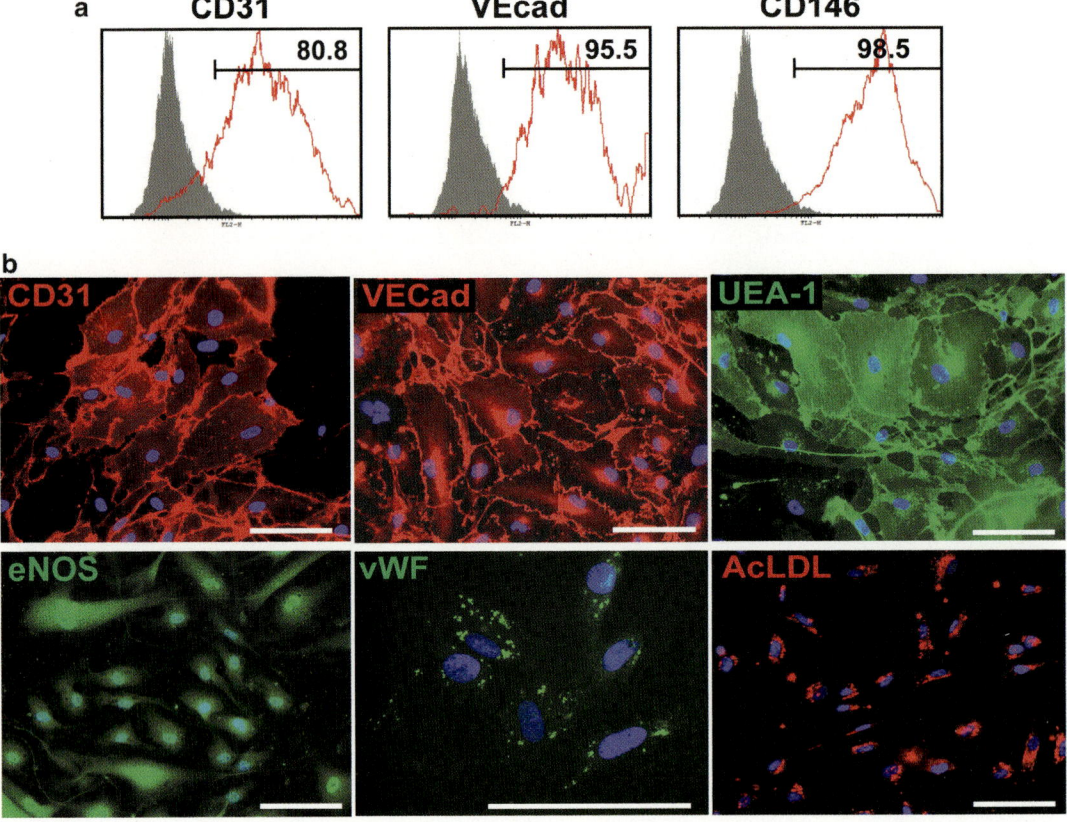

Fig. 2 EVC maturation toward ECs. Matured hPSC VEcad⁺ cells (**a**) are made up of a VEcad⁺CD31⁺CD146⁺ population (as demonstrated by flow cytometry) and (**b**) exhibit EC phenotypes such as membrane localization of CD31 and VEcad, lectin binding, cytoplasmic expression of eNOS, punctated vWF, and uptake of AcLDL, demonstrated by immunofluorescence. Nuclei in *blue*. Derivatives are demonstrated from hiPSC line BC1. Scale bars are 100 μm. Adapted from (10)

20. Add 1–2 ml EVC differentiation medium.

21. Repeat steps 18 and 19.

22. Replate cells on collagen IV-coated dishes at a cell density of 12,500 cells per cm^2.

23. Culture in 37 °C incubator at 5 % CO_2.

24. Change medium every other day.

25. Collect after 6 days for further analysis. Flow cytometry and immunofluorescence staining of these derivatives will reveal VEcad + CD31 + CD146+ cells with membrane localization of VEcad and CD31, lectin binding, cytoplasmic expression of eNOS, and von Willebrand factor (vWF) and uptake of acetylated low-density lipoprotein (AcLDL) (Fig. 2).

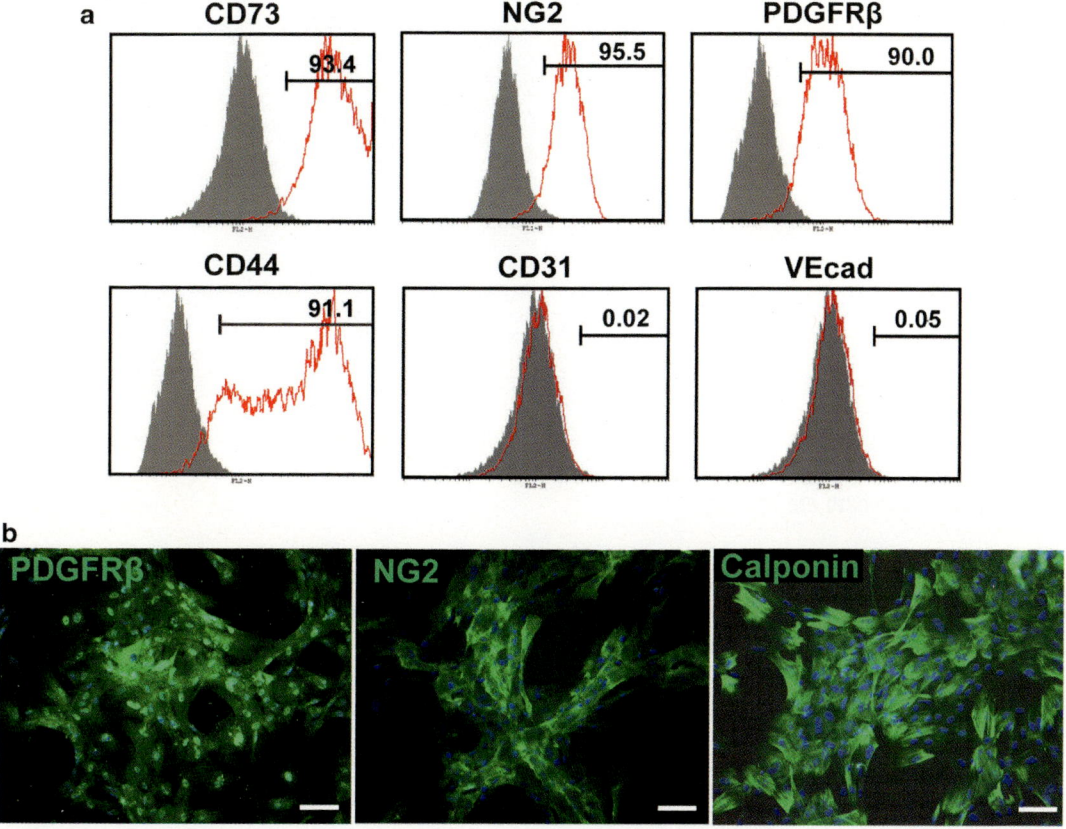

Fig. 3 EVC maturation toward pericytes. Matured hPSC pericytes are (**a**) enriched for CD73, NG2, PDGFRβ, and CD33; are depleted for CD31 and VEcad, as demonstrated by flow cytometry; and (**b**) exhibit appropriate localization of PDGFRβ, NG2, and calponin, as demonstrated by immunofluorescence. Derivatives are demonstrated from hiPSC line BC1. Scale bars are 100 μm. Adapted from (10)

3.3 Pericyte Maturation

1. Seed EVCs at a cell density of 12,500 per cm^2 onto tissue culture-treated wells in pericyte maturation medium.

2. Culture in 37 °C incubator at 5 % CO_2.

3. After 3 h, aspirate medium from wells and add fresh pericyte maturation medium.

4. Change medium every other day.

5. Collect after 6 days for further analysis. Flow cytometry and immunofluorescence staining of these derivatives will reveal enrichment in pericyte markers CD73, NG2, PDGFRβ, and CD44; depletion of EC markers VEcad and CD31; and appropriate localization of PDGFRβ, NG2 proteoglycan, and filamentous calponin expression (Fig. 3).

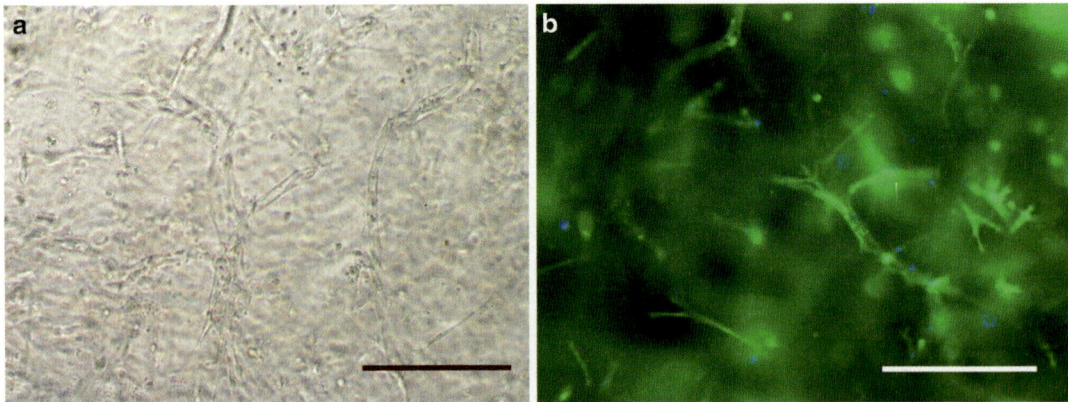

Fig. 4 EVC network formation. EVCs encapsulated with collagen hydrogels form multicellular networks after 3 days as assessed by (**a**) light microscopy and (**b**) immunofluorescence staining for phalloidin in *green* and nuclei in *blue*. Scale bars are 200 μm

3.4 Three-Dimensional Network Formation in Collagen Gels

1. Keep all reagents on ice during procedure to prevent premature collagen polymerization.

2. Coat wells of a 96-well plate with 100 μl PEI solution for 15 min.

3. Wash three times with PBS.

4. Coat wells with 100 μl glutaraldehyde solution for 30 min (see Note 4).

5. Wash three times with PBS.

6. Re-suspend approximately 800,000 EVCs in 200 μl M199 1× and place on ice (see Note 5).

7. In a separate Eppendorf tube, place 400 μl of M199 1×.

8. Add 40 μl M199 10×, and mix.

9. Add 350 μl collagen solution, and mix.

10. Add the entire cell suspension, and thoroughly mix.

11. Add 10 μl of 1 M NaOH to solution, and mix. The mixture should turn uniformly pink.

12. Add 56 μl of mixture to each well of coated 96-well plate (see Note 6).

13. Incubate at 37 °C for 30 min.

14. Add 100 μl of collagen gel culture medium.

15. Monitor network formation progress over 1–3 days via light or immunofluorescence microscopy as desired (Fig. 4).

4 Notes

1. Undifferentiated, healthy cultures of hPSCs are absolutely vital as starting populations. Cultures contaminated by differentiated cells will greatly affect differentiation potential.

2. Cell seeding density on day 6 is critical to ensure day-12 EVC phenotype. If too many cells are seeded, confluent cultures will inhibit VEcad expression.

3. If purity is not achieved on first MACS separation, it will be necessary to run another MACS separation with a fresh column on the sorted cells.

4. Steps 2–4 of the network formation procedure prevent contraction of the collagen gels.

5. This procedure will yield 1 mL of collagen hydrogels. Scale up or down for desired gel numbers.

6. Pipette collagen mix cautiously to avoid bubbles, which will preclude image acquisition.

References

1. Discher DE, Mooney DJ, Zandstra PW (2009) Growth factors, matrices, and forces combine and control stem cells. Science 324 (5935):1673–1677

2. Kusuma S, Gerecht S (2010) Engineering blood vessels using stem cells: innovative approaches to treat vascular disorders. Expert Rev Cardiovasc Ther 8(10): 1433–1445

3. Kusuma S, Gerecht S (2013) Recent progress in the use of induced pluripotent stem cells in vascular regeneration. Expert Rev Cardiovasc Ther 11(6):661–663

4. Stratman AN, Malotte KM, Mahan RD, Davis MJ, Davis GE (2009) Pericyte recruitment during vasculogenic tube assembly stimulates endothelial basement membrane matrix formation. Blood 114(24):5091–5101

5. Hanjaya-Putra D et al (2011) Controlled activation of morphogenesis to generate a functional human microvasculature in a synthetic matrix. Blood 118(3):804–815

6. Stegemann JP, Kaszuba SN, Rowe SL (2007) Review: advances in vascular tissue engineering using protein-based biomaterials. Tissue Eng 13(11):2601–2613

7. Zhang Z, Gupte MJ, Ma PX (2013) Biomaterials and stem cells for tissue engineering. Expert Opin Biol Ther 13(4):527–540

8. Couet F, Rajan N, Mantovani D (2007) Macromolecular biomaterials for scaffold-based vascular tissue engineering. Macromol Biosci 7(5):701–718

9. Chen Y-C et al (2012) Functional human vascular network generated in photocrosslinkable gelatin methacrylate hydrogels. Adv Funct Mater 22(10):2027–2039

10. Kusuma S et al (2013) Self-organized vascular networks from human pluripotent stem cells in a synthetic matrix. Proc Natl Acad Sci 110 (31):12601–12606

Methods in Molecular Biology (2014) 1202: 11–19
DOI 10.1007/7651_2014_75
© Springer Science+Business Media New York 2014
Published online: 28 March 2014

High-Throughput Cell Aggregate Culture for Stem Cell Chondrogenesis

Ahmad E. Abu-Hakmeh and Leo Q. Wan

Abstract

Cell aggregate culture is a widely used, reliable system for promoting chondrogenic differentiation of stem cells. A high-throughput cell pellet culture enables screening of various soluble factors for their effects on stem cell function and chondrogenesis. In this protocol, we report a platform that allows the formation of stem cell aggregates in a 96-well plate format. Specifically, stem cells are centrifuged to form high-density pellets, mimicking mesenchymal condensation. The cell aggregates can be differentiated into chondrocytes when cultured in chondrogenic medium for 4 weeks. Such a technique is compatible for high-throughput screening and can be very useful for optimizing conditions for cartilage tissue engineering.

Keywords: High-throughput, Stem cells, Chondrogenesis, Differentiation, Tissue engineering, Aggregate culture, Pellet

1 Introduction

Three-dimensional (3D) cell culture is critical for chondrogenic differentiation of stem cells indiscriminant of original tissue source (1). Chondrogenesis is thus achieved by one of two methods: (a) cell aggregate culture (2) or (b) cell encapsulation in a biocompatible material that is either inert or conducive to growth (3, 4). Such cultures are further improved by mimicking physicochemical cues of native articular cartilage with specific interest in selection and dosage of growth factors (5), seeding density (6), mechanical stimulation (7, 8), and oxygen tension (9, 10). Taken together, parameters are selected for their effects on morphology, matrix deposition, and cell–matrix interaction.

Traditional pellet culture involves centrifugation of cell-laden medium in ventilated microcentrifuge tubes and can be quite cumbersome to manage. Optimization of chondrogenic modulators can be facilitated by high-throughput screening. Here, we discuss a method for high-throughput cell pellet generation using a 96-well plate format. The procedure was first developed in 2005 (11) and has been adapted to 382-well plate culture (12) for more extensive growth factor and supplement dosage characterization. Typically, stem cells are cultured in monolayer to obtain the required total number for experimentation. Cells are then cleaved and reseeded at 200,000 to 500,000 cells per well and centrifuged (13). Cells are grown in a serum-free chemically defined medium that is replenished thrice weekly. Cultures are typically conducted for up to 4 weeks, followed by various end-point analyses. Here, we discuss pellet digestion and quantification of sulfated-glycosaminoglycans (s-GAG) as a means of quantifying chondrogenesis.

2   Materials

All solutions used for cell culture are prepared in a sterile cell culture hood. The culture hood and all reagents used therein are to be sprayed with 70 % ethanol to prevent contamination in cultures.

2.1   Stem Cell Monolayer Expansion Components

1. Human adipose-derived stem cells (hASCs).

2. hASC expansion medium: Remove 60 mL of DMEM-High Glucose (Gibco®; Carlsbad, CA) and store at 4 °C. In original bottle, add 50 mL Fetal Bovine Serum, 5 mL Penicillin/Streptomycin (PS; Sigma-Aldrich; St. Louis, MO), 5 mL Sodium Pyruvate (Na-Pyr; Gibco®), and 1 ng/mL bFGF (Invitrogen™; Carlsbad, CA) bringing volume to 500 mL. Store at 4 °C.

3. T-182 cm^2 flasks (Krackeler Scientific Inc.; Albany, NY).

4. Trypsin-EDTA, 0.25 % (Sigma-Aldrich). Store at −20 °C.

5. 15 mL polypropylene tubes (Krackeler Scientific, Inc.).

2.2   Cell Pellet Culture Components

1. Incomplete Chemically Defined Chondrogenic Medium (CDM$^-$; see **Note 1**): DMEM-HG: Remove 15 mL DMEM-HG and store at 4 °C. In original bottle, add 5 mL PS, 5 mL Na-Pyr, 5 mL ITS$^+$ Premix (BD^{TM}, Franklin Lakes, NJ), 0.1 μM Dexamethasone, and 40 μg/mL L-Proline (Sigma-Aldrich) bringing volume to 500 mL.

2. Complete chemically defined chondrogenic medium (CDM$^+$; see **Note 2**): Add 50 μg/mL ascorbate 2-phosphate (Sigma-Aldrich) and 10 ng/mL TGF-β3 (PeproTech, Rocky Hill, NJ) as fresh single-use aliquots to CDM$^-$.

3. Nontreated V-bottom 96-well plate (Sigma-Aldrich).

4. Universal well-plate lid with corner notch (Sigma-Aldrich).

2.3   Pellet Handling Prior to End-Point Analysis

1. Nontreated Polystyrene Culture dishes (Krackeler Scientific, Inc.).

2. Liquid Nitrogen.

2.4 DMMB Assay for Detection of s-GAG

1. Papain digestion buffer: Add 5 mM cysteine HCl, 5 mM EDTA, 100 mM Na_2HPO_4, and 125 µg/mL papain (Sigma-Aldrich) in dH_2O made fresh (*see* **Note 3**).

2. Synergy H1 microplate reader (BioTek Instruments; Winooski, VT).

3. Clear, flat-bottom 96-well plate (Sigma-Aldrich).

4. Glycine/NaCl solution: Add 0.3 % glycine and 0.24 % NaCl on bench top and dissolve with magnetic hot plate using a stirrer. Then add 0.01 M HCl (Sigma-Aldrich) under chemical hood.

5. DMMB solution: 16 µg 1,9-Dimethyl-Methylene Blue dye/mL (Sigma-Aldrich) in glycine/NaCl. Protect from light and store at 4 °C.

6. Phosphate buffered EDTA: 100 mM Na_2HPO_4 and 10 mM EDTA, pH 6.5.

7. Chondroitin sulfate standard: 1.0 mg/mL chondroitin sulfate A (CSA) from bovine trachea (Sigma-Aldrich) in phosphate-buffered EDTA as single-use aliquots. Store at −20 °C (*see* **Note 4**).

8. 30–300 µL Multichannel pipette (Eppendorf; Hamburg, Germany) (*see* **Note 5**).

9. 50 mL reagent reservoir (VWR, Radnor, PA).

2.5 Miscellaneous Culture Materials

1. DPBS.

2. 1.6 mL Polypropylene microcentrifuge tubes (Eppendorf).

3. Hemocytometer (Hausser Scientific Co; Horsham, PA).

4. Click Counter.

5. Trypan blue solution (MP Biomedicals, LLC; Santa Ana, CA).

6. Centrifuge 5804 (Eppendorf).

7. Isotemp™ Water Bath (Thermo Fisher Scientific™; Waltham, MA).

8. Phase Contrast Microscope.

9. 70 % Ethanol.

10. Cell Culture Hood.

11. Chemical Hood.

12. Magnetic hot plate and stirrer.

3 Methods

Carry out all procedures at room temperature unless otherwise specified. Cell culture is performed in a sterile culture hood. To optimize cell number with minimal sacrifice to differentiation capacity, hASCs are typically cultured up to passage four prior to downstream chondrogenic induction (14).

3.1 Thaw and Expansion of hASCs

1. Warm hASC medium in water bath at 37 °C.

2. Retrieve cryopreserved hASCs and rapidly thaw in water bath at 37 °C by shaking (*see* **Note 6**).

3. Transfer cell contents to 15 mL Polypropylene tube and centrifuge at $300 \times g$ for 5 min

4. Retrieve 15 mL tubes and observe cell aggregate.

5. Decant supernatant and resuspend pellet in 1 mL of hASC medium.

6. Transfer 20 µL of resuspended cells and 60 µL of Trypan Blue to 1.6 mL microcentrifuge tube.

7. Mix thoroughly and add 20 µL to hemocytometer.

8. Calculate total cell number.

9. Based on total cell number, calculate how many T-182 cm^2 flasks are needed for an approximate density of 6,000 cells/cm^2 (*see* **Note 7**).

10. Add 20 mL hASC medium to each T-182 cm^2 flask.

11. Add 6,000 cells/cm^2 to each flask.

12. Cap flask and gently rock north–south and east–west for 1 min (*see* **Note 8**).

13. Place in incubator at 37 °C and 5 % CO_2.

14. Change medium 3× a week until 80 % confluent (*see* **Note 9**).

3.2 Pellet Formation

1. Warm Trypsin-EDTA, CDM$^-$, and hASC media in water bath to 37 °C.

2. Thaw and add supplements to CDM$^-$ to make fresh CDM$^+$.

3. Retrieve T-182 cm^2 flasks from incubator and aspirate culture medium.

4. Wash flasks once with 6 mL of DPBS and aspirate.

5. Add 6 mL of Trypsin-EDTA to each flask and place in incubator at 37 ° C for 5 min (*see* **Note 10**).

6. Retrieve and neutralize trypsin with 6 mL of hASC medium.

7. Transfer contents into 15 mL tube and centrifuge at $300 \times g$ for 5 min.

8. Retrieve 15 mL tubes, observe cell aggregate, decant supernatant, and resuspend in 1 mL of CDM$^+$ medium (*see* **Note 11**).

9. Transfer 20 µL of cell-laden medium and 60 µL of Trypan blue to a microcentrifuge tube,

10. Mix thoroughly by pipetting up and down and load 20 µL to hemocytometer.

11. Calculate total cell number and add CDM$^+$ medium to make density 2.5×10^6 cells/mL.

12. Add 200 μL of cell-laden CDM$^+$ medium per well on 96-well plate (*see* **Notes 12** and **13**).

13. Centrifuge at 300 × g for 5 min (*see* **Note 14**).

14. Place plates in incubator at 37 °C and 5 % CO_2 undisturbed for 2 days (*see* **Note 15**).

15. Change medium 3× a week (*see* **Note 16**).

3.3 Digestion of Aggregates for End-Point Analysis (See Note 17)

1. Using a pipette, aspirate medium from 96-well plate.

2. Wash cell aggregates with PBS once and aspirate.

3. Using a multichannel pipette directly add 200 μL of papain digestion buffer to each well.

4. Digest samples by placing 96-well plate in water bath at 60 °C for 18 h (*see* **Note 18**).

5. Store contents at −20 °C or proceed with downstream assay.

3.4 DMMB Assay for Quantification of s-GAG

DMMB is a cationic dye that binds to s-GAG. Colorimetric assaying with DMMB blue can be used to quantitate prominent s-GAG in cartilage and other tissues (15). While this method is useful for quantifying s-GAG, the DMMB–GAG complex is very unstable. The spectrophotometer should be calibrated before the dye is added to the samples. Readings should be taken immediately after adding DMMB to the 96-well plate. All steps requiring aspiration should refer to the steps outlined in **Note 16**.

1. Thaw frozen CSA standard and dilute to 50 μg/mL in dH_2O to make CSA working solution.

2. Using a clear, flat-bottom 96-well plate, make working dilutions of 0–50 μg/mL as indicated in Table 1.

3. Turn on microplate reader (*see* **Note 19**).

4. Set and calibrate microplate reader to measure absorbance at 540 and 595 nm sequentially (*see* **Note 20**).

5. Using the multichannel pipette, add 250 μL of DMMB solution to standard wells and mix by pipetting up and down (*see* **Note 21**).

6. Measure absorbance at 540 and 595 nm (*see* **Note 22**).

7. Develop standard curve and determine linear region.

Table 1
Preparation of chondroitin sulfate A (CSA) standards

Concentration CSA (μg/mL)	0	5	10	20	30	40	50
μL of CSA standard (50 μg/mL)	0	5	10	20	30	40	50
μL of dH_2O	50	45	40	30	20	10	0

8. For measurement of samples; thaw sample digest stored at $-20\ ^{\circ}$C.

9. First transfer 50 µL of an aliquot of a single dissolved sample in a well.

10. Add 250 µL of DMMB dye and measure at absorbance (*see* **Note 23**).

11. Using a multichannel pipette, proceed to transfer 50 µL aliquots of all remaining pellet samples to the 96-well plate at the determined dilution factor.

12. Add DMMB dye and measure absorbance.

13. Calculate concentration of s-GAG per sample by linearly interpolating absorbance measurements using the CSA standard curve.

4   Notes

1. We choose to use chemically defined medium for these experiments. Various other studies have effectively used serum-supplemented chondrogenic medium (16).

2. For best results, make aliquots of AA2P and TGF-β3 (typically of 1,000 times the final desired concentration) and store at 4 °C and $-20\ ^{\circ}$C, respectively. If a TGF-β3 aliquot has more than a single use, then do not refreeze in $-20\ ^{\circ}$C. Rather, store at 4 °C until next use.

3. Papain is unstable under acidic conditions below pH values of 2.8. EDTA and cysteine are added as stabilizing agents (17).

4. While $-20\ ^{\circ}$C is sufficient for long term storage, we have found that aliquots stored at $-80\ ^{\circ}$C are stable for at least 2 years.

5. Using an 8-channel pipette is sufficient for making the standard curve; however, in an experiment with many samples, a 12-channel pipette is preferred due to DMMB-GAG complex instability and the necessity of rapid assay execution.

6. This should take 1–2 min. To avoid potential contamination, refrain from submerging the cap of the cryo-vial under water.

7. For convenience, we freeze cells at a density of 1.1×10^6 cells/ mL such that each vial would be expanded in an independent T-182 cm^2 flask. 1.1×10^6 cells per vial corresponds to approximately 6,000 cells/cm^2.

8. This is done to promote homogenous cell distribution. Confirm this using a phase contrast microscope prior to storing in the incubator. Insufficient disaggregation of pellet formed in Sections 3.1–3.3 will manifest as clusters of floating cells at this stage and ultimately as a flask with heterogeneous cell distribution.

9. Using hASCs, a confluency of 80 % yields approximately 44,000 cells/cm^2.

10. Used to cleave cells from flask; volume used is optimized for size of flask and final cell density. Over-confluent cells may need 7 mL of Trypsin.

11. Cells resuspended in 1 mL of medium will be at a higher density than desired for pellet inoculation. A secondary dilution following cell counting is needed.

12. Using CDM$^+$, we find that a density of 5×10^5 cells/well promotes a pellet size useful for end-point analysis such as sectioning for staining. Due to enhanced proliferation in serum-supplemented medium, a density of 2.5×10^5 cells is sufficient.

13. To promote uniform cell aggregates, partition cell-laden medium into multiple and polypropylene tubes and mix periodically in between adding to 96-well plate.

14. Fill unused wells with sterile DPBS for balance during centrifugation. Also it is useful to have a blank 96-well plate with water in all wells to balance any odd number of experimental plates.

15. Pellets are easily disturbed and take a couple of days to aggregate properly. Early disruption could result in partial-to-complete disaggregation.

16. Although pellets stabilize after the first 2 days, all precautions should be taken to promote pellet homogeneity for the duration of the culture. Thus, we find that removing 50 µL at a time is optimal. Removing volumes in excess of 100 µL may cause disaggregation or distortion into a more elliptical-shaped pellet. An emphasis is put on manually removing medium with a pipette rather than aspiration as this can also compromise pellet morphology or result in accidental aspiration.

17. This procedure is specific for End-Point analysis and sample preparation for chemical quantification such as DNA, s-GAG, and collagen content. For RNA extraction, the user should follow the protocol as provided by the employed kit or snap freeze in liquid nitrogen and store at -80 °C until eventual use. For sectioning tissues, the user should fix and embed aggregates in paraffin or OCT.

18. Volume is optimized based on seeding density of aggregates per pellet. If unsure about final aggregate cell number or chemical content, it is superior to digest pellets in a lesser volume of papain (e.g. 100 µL) and subsequently add more digestion medium as needed to fit the linear range of standard curves in downstream assays. In the case of over-diluting samples, a microassay standard curve of lower concentration CSA aliquots is needed. The linearity may be different at significantly

low concentrations and cannot be assumed to fit a standard curve generated for higher CSA concentrations.

19. If using a nonmonochromatic microplate reader, power on 15 min prior to running the DMMB assay.

20. Sequentially measured wavelengths in a single test are preferred due to the instability of the DMMB-GAG complex.

21. Pipette carefully to avoid generating bubbles. Bubbles interfere with absorbance readings. It is best to add this dye on a bench top directly next to the microplate reader. The stability of the DMMB-GAG complex may be compromised in as little as 5 min.

22. Absorbance increases with concentration of s-GAG at 540 nm and decreases at 595 nm. Plotting the difference in the optical densities gives a more sensitive standard curve. Estimating s-GAG concentration should be calculated based on absorbances reflecting the difference in optical densities at the experimental CSA concentrations.

23. This reading is used to see if the sample falls within the standard curve. If the absorbance is above the linear limit, then the sample must be further digested in papain solution. Set any dilution such that the sample concentration is towards the middle of the standard curve.

Acknowledgment

LQW wants to thank the National Science Foundation for funding support.

References

1. Gadjanski I, Spiller K, Vunjak-Novakovic G (2012) Time-dependent processes in stem cell-based tissue engineering of articular cartilage. Stem Cell Rev 8(3):863–881

2. Ahrens PB, Solursh M, Reiter RS (1977) Stage-related capacity for limb chondrogenesis in cell culture. Dev Biol 60(1):69–82

3. Awad HA, Wickham MQ, Leddy HA et al (2004) Chondrogenic differentiation of adipose-derived adult stem cells in agarose, alginate, and gelatin scaffolds. Biomaterials 25 (16):3211–3222

4. Kim IL, Khetan S, Baker BM et al (2013) Fibrous hyaluronic acid hydrogels that direct MSC chondrogenesis through mechanical and adhesive cues. Biomaterials 34(22):5571–5580

5. Jakob M, Démarteau O, Schäfer D et al (2001) Specific growth factors during the expansion and redifferentiation of adult human articular chondrocytes enhance chondrogenesis and cartilaginous tissue formation in vitro. J Cell Biochem 81(2):368–377

6. Hui TY, Cheung KMC, Cheung WL et al (2008) In vitro chondrogenic differentiation of human mesenchymal stem cells in collagen microspheres: Influence of cell seeding density and collagen concentration. Biomaterials 29(22):3201–3212

7. Graff RD, Lazarowski ER, Banes AJ et al (2000) ATP release by mechanically loaded porcine chondrons in pellet culture. Arthritis Rheum 43(7):1571–1579

8. Angele P, Schumann D, Angele M et al (2004) Cyclic, mechanical compression enhances chondrogenesis of mesenchymal progenitor cells in tissue engineering scaffolds. Biorheology 41(3):335–346

9. Meyer EG, Buckley CT, Thorpe SD et al (2010) Low oxygen tension is a more potent promoter of chondrogenic differentiation than dynamic compression. J Biomech 43 (13):2516–2523

10. Malladi P, Xu Y, Chiou M et al (2006) Effect of reduced oxygen tension on chondrogenesis and osteogenesis in adipose-derived mesenchymal cells. Am J Physiol Cell Physiol 290(4): C1139–C1146

11. Penick K, Solchaga L, Welter J (2005) High-throughput aggregate culture system to assess the chondrogenic potential of mesenchymal stem cells. Biotechniques 39(5):687–691

12. Huang AH, Motlekar NA, Stein A et al (2008) High-throughput screening for modulators of mesenchymal stem cell chondrogenesis. Ann Biomed Eng 36(11):1909–1921

13. Merceron C, Portron S, Vignes-Colombeix C et al (2012) Pharmacological modulation of human mesenchymal stem cell chondrogenesis by a chemically oversulfated polysaccharide of marine origin: potential application to cartilage regenerative medicine. Stem Cells 30 (3):471–480

14. Guilak F, Lott KE, Awad HA et al (2006) Clonal analysis of the differentiation potential of human adipose-derived adult stem cells. J Cell Physiol 206(1):229–237

15. Farndale RW, Buttle DJ, Barrett AJ (1986) Improved quantitation and discrimination of sulphated glycosaminoglycans by use of dimethylmethylene blue. Biochim Biophys Acta 883(2):173–177

16. Wan LQ, Jiang J, Miller DE et al (2011) Matrix deposition modulates the viscoelastic shear properties of hydrogel-based cartilage grafts. Tissue Eng Part A 17 (7–8):1111–1122

17. Arnon R (1970) Papain. Methods Enzymol 19:226–244

Methods in Molecular Biology (2014) 1202: 21–27
DOI 10.1007/7651_2013_36
© Springer Science+Business Media New York 2013
Published online: 24 October 2013

Microfluidic Device to Culture 3D In Vitro Human Capillary Networks

Monica L. Moya, Luis F. Alonzo, and Steven C. George

Abstract

Models that aim to recapitulate the dynamic *in vivo* features of the microcirculation are crucial for studying vascularization. Cells *in vivo* respond not only to biochemical cues (e.g., growth factor gradients) but also sense mechanical cues (e.g., interstitial flow, vessel perfusion). Integrating the response of cells, the stroma, and the circulation in a dynamic 3D setting will create an environment suitable for the exploration of many fundamental vascularization processes. Here in this chapter, we describe an *in vivo*-inspired microenvironment that is conducive to the development of perfused human capillaries.

Keywords: Microfluidic platform, Microenviroment, Perfused capillary networks, Vasculogenesis, Tissue engineering

1 Introduction

Understanding the mechanisms that regulate vascularization is essential for the development of tissue engineering strategies. Our current understanding of the vascularization process has primarily relied on 2D and 3D co-culture *in vitro* models as well as *in vivo* animal models. From these model systems we have gained significant knowledge about fundamental cellular interactions between the primary components of vascular structures, endothelial cells, and the surrounding stromal cells (i.e., fibroblasts, pericytes, smooth muscle cells) which guide their development and maturation via paracrine signaling and physical contact (1, 2). Many of these systems, however, lack the complexity needed to mimic the dynamic *in vivo* microenvironment such as interstitial flow and vessel perfusion. To address these limitations, various research groups have explored the use of microfluidic technologies in order to create controllable, dynamic *in vivo*-like environments (3–8). These platforms have shown the potential of recreating relevant environmental cues (e.g., cell–cell communications, growth factor gradients, interstitial flow) while maintaining flexible and convenient culture conditions (e.g., high-throughput, real-time imaging).

Here we describe a microfluidic-based system that allows for real-time visualization of perfused human capillary networks in a dynamic microenvironment that includes interstitial flow. In this system, endothelial cells and stromal cells are allowed to self-assemble in a 3D matrix that is conducive for the development of perfused human capillaries, achieving flows within the physiological range (3). This platform can be used to study the effects of interstitial flow (5), heterotypic and homotypic cell interactions, and growth factors during the development of *in vitro* human capillary networks. In addition, this platform is easily adapted to include tissue-specific function (e.g., cardiac muscle) as well as models of disease that involve the microcirculation.

2 Materials

2.1 Polydimethyl-siloxane Microfluidic Devices

1. Prepare polydimethylsiloxane (PDMS) by mixing the cross-linker, Sylgard 184, with the curing agent (Dow Corning) at a ratio of 10:1. For example, 10 g of PDMS are cured with 1 g of curing agent. Ensure that the curing agent and cross-linker are well mixed.

2. Degas mixture in a vacuum chamber for 30 min to remove excess air bubbles from solution.

3. After 30 min, solution should be essentially bubble free and can then be poured onto an SU-8-patterned microfluidic casting mold (approximately 20–25 g). Note that details on how to create an SU-8 mold are not included in this report as this can be found in a variety of additional references (9, 10).

4. Place the SU-8 mold coated with the PDMS solution in the vacuum chamber for an additional 15 min to remove any further air bubbles.

5. Place the mold with PDMS into an oven at 65 °C, and allow to cure for at least 8 h (and up to 2 days) prior to removing from mold.

2.2 Assembly of Sealed Devices

1. Carefully remove the PDMS device from the mold using a surgical blade to cut around the perimeter of the device. Carefully lift off the mold (*see Note 1*).

2. Using a 16G needle, punch into the device the inlet and outlet of the tissue chambers and the microfluidic lines. Using a nitrogen air gun, blow away any excess debris from the device surface (*see Note 2*).

3. Plasma treat for 3.5 min all three pieces which make up the microfluidic device: glass slide/cover glass, a separate thin sheet of PDMS (thickness 500–750 μm), and PDMS piece with negative replica of mold design.

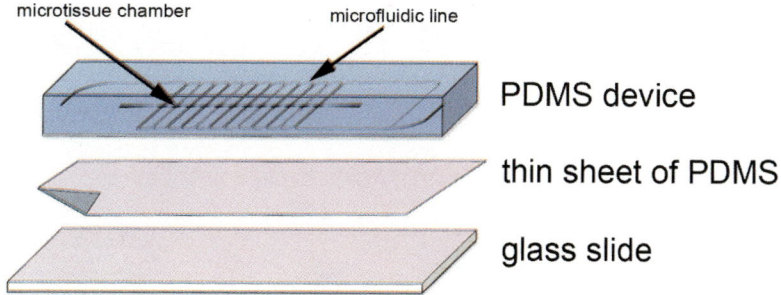

Fig. 1 Assembly of microfluidic device for *in vitro*-perfused human capillaries. Plasma-treated surfaces (indicated in *pink* color) of PDMS device, sheet of PDMS, and glass slide are bonded together and baked in 120 °C oven for 15 min to obtain PDMS-enclosed chambers

4. Bond all three plasma-treated pieces to each other quickly (<90 s) to seal the microfluidic device. The PDMS piece representing the negative replica of the mold design should be facing up and be bonded to the thin sheet of PDMS. The glass slide-treated surface is then bonded to thin sheet of PDMS (*see Note 3*) (Fig. 1).

5. Place bonded device in 120 °C oven for 15 min.

6. Bond glass vials (i.e., media reservoirs) over inlet/outlet holes of assembled microfluidic device using a mixture of PDMS. To prevent PDMS seeping into inlet/outlet holes, place pipette tips on all inlet/outlet holes of assembled microfluidic device.

7. Dip the bottom of a precut glass vial into the PDMS mixture, and place through the pipette tip onto the top surface of the device.

8. Place completely assembled device in 65 °C oven overnight to cure.

9. Sterilize assembled PDMS device (Fig. 2) in autoclave.

2.3 Cell Preparation

1. Trypsinize, collect, and count cells needed for experiment.

2. Reconstitute cells at desired cell density (see chart), and combine. Cells can then be spun down together. Note that cord blood endothelial colony-forming cell-derived endothelial cells (ECFC-EC) and normal human lung fibroblasts (NHLF) work well for this assay. However, other endothelial cell sources (e.g., human umbilical vein endothelial cells, HUVEC) and stromal cell sources (e.g., dermal fibroblast) have also been successfully substituted (11–13).

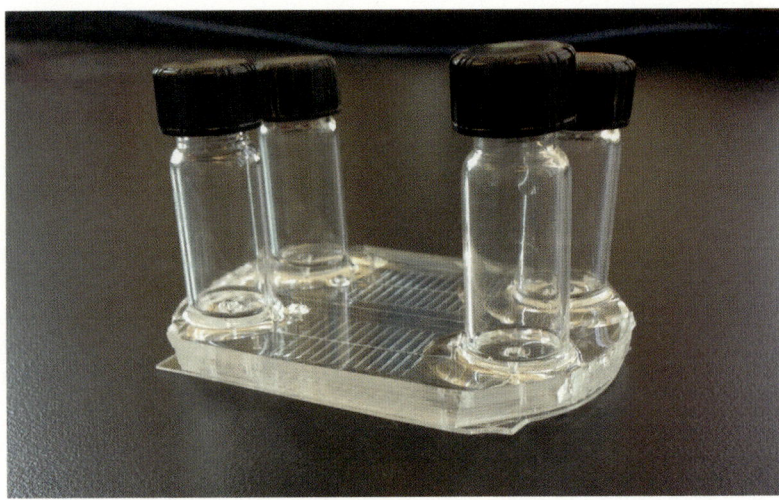

Fig. 2 Macroscopic view of assembled microfluidic device with glass vial reservoirs attached to the PDMS chamber bonded to a thin layer of PDMS and a glass slide. Also visible are the microfluidic lines that will contain cell culture media or fibrin and cells

Experiment	Cell density
Vasculogenesis model	
ECFC-ECs	2.5×10^6 cells/ml
NHLFs	5×10^6 cells/ml
Tumor vasculogenesis model	
SW620s	160,000 cells/ml
ECFC-ECs	2.5×10^6 cells/ml
NHLFs	5×10^6 cells/ml

2.4 Hydrogel Preparation

1. Prepare fibrinogen (Sigma) by reconstituting in Dulbecco's phosphate-buffered saline (DPBS) to a final concentration of 10 mg/ml.

2. Allow fibrinogen solution to warm, and dissolve by placing in 37 °C water bath for a minimum of 15 min.

3. Once fibrinogen solution is fully dissolved, the solution should then be sterile filtered using a 0.22 μm syringe filter.

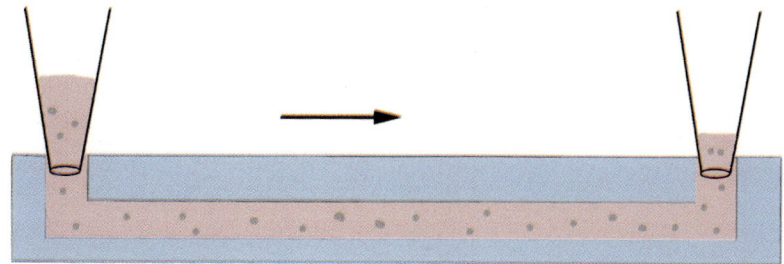

Fig. 3 Side-view schematic of loading tissue chambers. Prior to polymerization, the fibrinogen–cell mixture is micropipetted into the inlet of the tissue chamber using steady even pressure

3   Methods

3.1   Loading Device

1. Remove devices from autoclave, and allow to cool at room temperature before proceeding to load with hydrogel and cells.

2. Resuspend cells in prepared fibrinogen according to desired volume and cell density (*see Note 4*).

3. Pipette 30 μl of cell–fibrinogen mixture into a microcentrifuge tube containing 1.8 μl of 50 U/ml thrombin, and mix thoroughly while avoiding the creation of air bubbles (*see Note 5*).

4. Using a filtered pipette tip, quickly (<3 min) collect a minimum of 25 μl of mixture and insert into the inlet of the tissue channel applying steady and even pressure to avoid leakage of hydrogel into adjacent fluidic channels. Keep applying a steady pressure until the mixture reaches the outlet of the tissue chamber (Fig. 3) (*see Note 6*).

5. Incubate device in 37 °C humidified incubator for a minimum of 30 min and up to 1 h to ensure full fibrin polymerization.

3.2   Introducing Media into Fluidic Lines

1. Insert 200 μl of fully supplemented EGM-2 (Lonza) into the inlet of one adjacent microfluidic channel applying steady, even pressure until the media reaches the outlet of the microfluidic channel forming a small bubble of media.

2. For the other microfluidic line, insert into the outlet a small piece of tygon tubing (0.02″ ID × 0.06″ OD) to connect or couple the two microfluidic lines. Repeat previous step with the second channel ensuring that the media enters and forms a small droplet of fluid at the end of the microfluidic line. Alternatively, the microfluidic lines may remain uncoupled, if desired, by not employing the small plastic tubing.

3. Connect fluid-filled tubing with the outlet of the first channel, and avoid introducing any air gaps (*see Note 7*).

4. Once connected, fill reservoirs with media. Set a height difference of 10 mm between the two reservoirs to allow media to follow through the fluidic channels as well as across the tissue chambers. Note that a larger or a smaller height difference can be employed to produce higher or lower flows, respectively, if desired.

5. Place assembled device into a 20 % O_2 incubator overnight.

3.3 Maintenance of Culture Within the Microfluidic Device

1. The day after seeding the device, replace the media with EGM-2 without VEGF and bFGF. Device can then be moved to a 5 % O_2 incubator for the remainder of the study, if desired (*see Note 8*).

2. Continue to level the fluid in the reservoir every other day to maintain the 10 mm (or other) height difference.

3. After 7 days of culture, reverse the direction of the interstitial flow across the tissue chamber by reversing the pressure head in the reservoirs. This encourages vessels to grow towards both pores on either side and form connections with the microfluidic channels.

4   Notes

1. For easy removal of the PDMS mold from wafer mold, silanize the wafer prior to casting.

2. If bubbles occur between the layers of PDMS, use a flat-edge surface to press bubbles away from chambers or microfluidic lines. Bubbles in between layers away from chambers or microfluidic line are not a problem as they do not interfere with the loading of the device.

3. Use scotch tape on the surfaces of the device to remove stubborn debris from the PDMS device.

4. Aliquot cell mixtures in smaller sets for loading to avoid premature polymerization of hydrogel. For example if loading eight samples, split cells into two sets of four loadings.

5. Fibrinogen–cell mixture will begin polymerizing quickly upon adding thrombin, so perform this step quickly to avoid damaging the gel or creating uneven polymerization.

6. To avoid introducing bubbles into cell–fibrinogen–thrombin mixture, change tips in between mixing the solution and loading the solution into the device.

7. When connecting tubing and microfluidic channels make sure to touch the liquid-formed droplet at the microfluidic line

outlet to the small droplet formed at the tip of the tubing so as to always maintain fluid–fluid contact and avoid introducing air bubbles.

8. To minimize evaporation of media from the reservoirs of the device while in the incubator, create a secondary humidity chamber by placing the microfluidic devices into a box with a non-airtight lid containing a reservoir of water of about 20 ml.

References

1. Newman AC, Nakatsu MN, Chou W, Gershon PD, Hughes CCW (2011) The requirement for fibroblasts in angiogenesis: fibroblast-derived matrix proteins are essential for endothelial cell lumen formation. Mol Biol Cell 22 (20):3791–3800. doi:10.1091/mbc.E11-05-0393

2. Montesano R, Pepper MS, Orci L (1993) Paracrine induction of angiogenesis *in vitro* by Swiss 3T3 fibroblasts. J Cell Sci 105 (4):1013–1024

3. Moya ML, Hsu YH, Lee AP, Hughes CC, George SC (2013) In vitro perfused human capillary networks. Tissue Eng Part C Methods 19(9):730–737. doi:10.1089/ten.TEC.2012.0430

4. Vickerman V, Blundo J, Chung S, Kamm R (2008) Design, fabrication and implementation of a novel multi-parameter control microfluidic platform for three-dimensional cell culture and real-time imaging. Lab Chip 8(9):1468–1477. doi:10.1039/b802395f

5. Hsu YH, Moya ML, Abiri P, Hughes CC, George SC, Lee AP (2013) Full range physiological mass transport control in 3D tissue cultures. Lab Chip 13(1):81–89. doi:10.1039/c2lc40787f

6. Hsu Y-H, Moya ML, Hughes CCW, George SC, Lee AP (2013) A microfluidic platform for generating large-scale nearly identical human microphysiological vascularized tissue arrays. Lab Chip 13(15):2990–2998. doi:10.1039/c3lc50424g

7. Kim S, Lee H, Chung M, Jeon NL (2013) Engineering of functional, perfusable 3D microvascular networks on a chip. Lab Chip 13(8):1489–1500. doi:10.1039/c3lc41320a

8. Song JW, Bazou D, Munn LL (2012) Anastomosis of endothelial sprouts forms new vessels in a tissue analogue of angiogenesis. Integr Biol (Camb) 4(8):857–862. doi:10.1039/c2ib20061a

9. Whitesides GM, Stroock AD (2001) Flexible methods for microfluidics. Phys Today 54 (6):42–48. doi:10.1063/1.1387591

10. Duffy DC, McDonald JC, Schueller OJA, Whitesides GM (1998) Rapid prototyping of microfluidic systems in poly(dimethylsiloxane). Anal Chem 70(23):4974–4984. doi:10.1021/Ac980656z

11. Chen X, Aledia AS, Ghajar CM, Griffith CK, Putnam AJ, Hughes CC, George SC (2009) Prevascularization of a fibrin-based tissue construct accelerates the formation of functional anastomosis with host vasculature. Tissue Eng Part A 15(6):1363–1371. doi:10.1089/ten.tea.2008.0314

12. Ghajar CM, Blevins KS, Hughes CC, George SC, Putnam AJ (2006) Mesenchymal stem cells enhance angiogenesis in mechanically viable prevascularized tissues via early matrix metalloproteinase upregulation. Tissue Eng 12 (10):2875–2888. doi:10.1089/ten.2006.12.2875

13. Chen XF, Aledia AS, Popson SA, Him L, Hughes CCW, George SC (2010) Rapid Anastomosis of Endothelial Progenitor Cell-Derived Vessels with Host Vasculature Is Promoted by a High Density of Cotransplanted Fibroblasts. Tissue Eng Part A 16(2):585–594. doi:10.1089/Ten.Tea.2009.0491

Methods in Molecular Biology (2014) 1202: 29–36
DOI 10.1007/7651_2013_62
© Springer Science+Business Media New York 2013
Published online: 7 February 2014

Multifunction Co-culture Model for Evaluating Cell–Cell Interactions

Danielle R. Bogdanowicz and Helen H. Lu

Abstract

Interactions within the same cell population (homotypic) and between different cell types (heterotypic) are essential for tissue development, repair, and homeostasis. To elucidate the underlying mechanisms of these cellular interactions, co-culture models have been used extensively to investigate the role of cell–cell physical contact, autocrine and/or paracrine interactions on cell function, as well as stem cell differentiation. Specifically, the mixed co-culture model is often optimal for interpreting the effects of cell–cell contact on cellular behavior in vitro, while indirect co-culture can be used to study the effects of paracrine signaling on cell reactions. Additionally, cell–cell contact can be controlled by establishing physical barriers, which are used to regulate spatial and temporal cell distribution patterns in co-culture. In this chapter, we describe a method for forming a removable permeable divider for temporally and spatially controlling cellular interactions. This model can be used to study the impact of both cell–cell contact and paracrine signaling on the behavior of the mixed population as a whole and on the response of each subpopulation of cells in co-culture.

Keywords: Co-culture, Stem cells, Live cell tracking, Hydrogel divider, In vitro

1 Introduction

The maintenance, remodeling, and repair of biological tissue rely on the interaction of cells with other cell types and with their surrounding environment (1, 2). Specific to regenerative medicine, there is a growing interest in the nature and ramifications of the interactions between mesenchymal stem cells and resident cell populations in a given tissue or at the repair site, as stem cells represent a clinically feasible cell source for tissue engineering and regenerative medicine because of their trophic capabilities, as well as their potential to differentiate toward a number of mesenchymal lineages (3–5). The mechanisms behind the interactions between these cell types are, however, not well understood (6). For this reason, the development of in vitro model systems capable of modulating these interactions, which allow researchers to investigate the role of cellular communication in tissue formation and regeneration, is much needed.

To date, the majority of current co-culture models can be categorized into two types: those which examine the effects of direct cell–cell contact on co-cultured cell populations (7–12) and those which focus on paracrine signaling and response to soluble signaling factors (13–17). These studies can be conducted at relatively macro- (18, 20, 21) or microscales (22–25). The simplest model for examining the effects of cell–cell contact involves the mixing of two cell types and subsequently seeding a monolayer of this mixed population. This model allows for control over the extent of heterotypic and homotypic interactions through alteration of the seeding densities of each cell type and relative seeding ratio of the subpopulations (26). However, it is difficult to directly determine the relative contributions of each cell type to any observed effects of co-culture; thus, these studies are often accompanied in parallel by conditioned media experiments. In studies in which the effects of paracrine signaling are of interest, a segregated co-culture system can be utilized. This is done by first growing individual cell populations separately, followed by culturing both cell types in the same environment, in the absence of direct cell contact. This can be achieved either through the use of conditioned media studies or by forming a physical barrier or membrane between cell types that permits the exchange of soluble factors, while still preventing cell–cell contact. One advantage of this system is that the analysis of the individual response of each subpopulation of cells due to co-culture can be readily determined (26).

In the methods outlined below we describe a model which allows for the study of the effects of both cell–cell contact and paracrine signaling on subpopulations of cells in co-culture. Moreover, this model exerts spatial and temporal control over heterotypic cellular interactions through the development of a removable permeable hydrogel barrier. This in vitro model system permits the live tracking of populations throughout in vitro culture and allows for the study of individual cell subpopulation response to co-culture (20).

2   Materials

2.1 Hydrogel Divider Components

Phosphate-buffered saline (PBS, Sigma Chemical, St. Louis, MO, USA): One packet PBS powder into 1,000 mL deionized water (see Note 1).

Low gelling temperature agarose (Sigma Chemical, St. Louis, MO, USA).

Hydrogel solution: 4 % weight per volume (w/v) low gelling temperature agarose powder in 1× sterile PBS.

2.2 Live Cell Tracker Reagent Components

Fully supplemented (F/S) culture media: Dulbecco's modified essential medium (DMEM) with 10 % fetal bovine serum, 1 % nonessential amino acids, 1 % antibiotics (see Notes 2 and 3).

Serum-free culture media: DMEM with 1 % nonessential amino acids, 1 % antibiotics.

Vybrant™ DiO cell-labeling solution (Molecular Probes, Eugene, OR, USA) (*see* **Note 4**).

Vybrant™ DiD cell-labeling solution (Molecular Probes, Eugene, OR, USA).

3 Methods

Carry out all procedures at room temperature under sterile conditions unless otherwise specified.

3.1 Hydrogel Divider Formation and Placement

1. Sterile filter 1× PBS into a sterile container (*see* **Notes 5** and **6**).
2. Mix agarose powder into 1× PBS to achieve a final agarose concentration of 4 % w/v.
3. Autoclave for 25 min using a liquid cycle to sterilize.
4. Cast agarose gel by pipetting 2 mL of agarose solution into a 3.8 cm^2 tissue culture plastic well and allow cooling (*see* **Note 7**).
5. Create a 5 mm hydrogel divider by cutting a 5 mm-wide strip along the center of the well using a scalpel. Slide the blade along the outer boundary of the well afterwards to separate the strip from the well.
6. Gently remove the 5 mm strip from the well and transfer to a new well plate.
7. Coat the bottom surface and sides of the divider with unsolidified agarose solution, using a sterile spatula or a pipette tip.
8. Place the coated divider into a new 3.8 cm^2 well, and allow the agarose glue to dry and the divider to adhere to the sides and bottom of the tissue culture plastic surface (Fig. 1).
9. Parafilm seal the well plate containing the divider using parafilm and store at 37 °C until cell seeding to prevent gel dehydration.
10. Repeat steps 1–10 for each sample.

3.2 Cell Membrane Staining

All solutions used should be warmed to 37 °C before use.

1. Remove stem cell (Cell Type A) culture plates from cell culture incubator (*see* **Note 8**).
2. Establish a cell suspension via traditional trypsinization methods.
3. Remove cell suspension from the tissue culture plate and transfer the suspension into a sterile polypropylene centrifuge tube.
4. Centrifuge cells at 1,500 rpm for 12 min at room temperature to form a cell pellet.
5. Remove supernatant (*see* **Note 9**).

Fig. 1 Hydrogel barrier formed following gel setting and removal of excess hydrogel

6. Resuspend cell pellet in a minimum volume of F/S culture media.

7. Count cells and assess viability (*see* **Note 10**).

8. Centrifuge cells at 1,500 rpm for 10 min at room temperature.

9. Resuspend cells in serum-free culture media at a density of 1×10^6/mL.

Steps 15–20 should be performed in the dark, preferably (Fig. 2).

10. Add 5 μL DiD Vybrant™ Cell-Labeling Solution (1 mM in solvent) per mL of cell suspension. Mix well by gently pipetting up and down.

11. Incubate the cell suspension for 20 min in the dark at 37 °C (*see* **Notes 11** and **12**).

12. Centrifuge the cell suspension at 1,500 rpm for 5 min, preferably at 37 °C.

13. Remove supernatant, and gently resuspend cells in warm serum-free culture media.

14. Repeat steps 17 and 18 two more times.

15. Resuspend cells in F/S culture media at a density of 1×10^5/mL.

16. Incubate in the dark at 37 °C until ready for seeding (*see* **Note 13**).

17. Repeat steps 1–16 for second cell type (Cell Type B), using DiO Vybrant™ Cell-Labeling Solution (1 mM in solvent) instead of DiD Vybrant™ Cell-Labeling Solution in step 10.

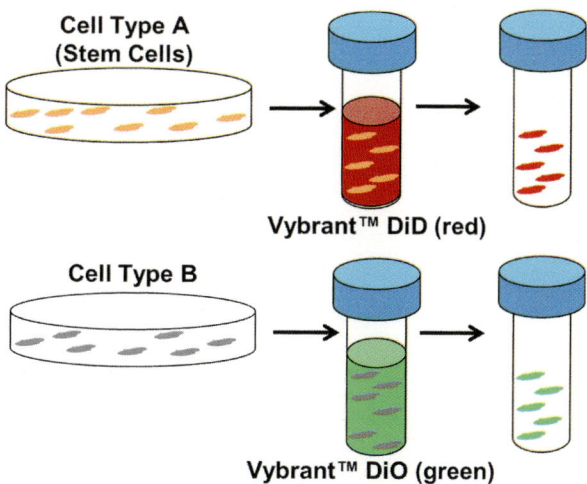

Fig. 2 Schematic of membrane staining using Vybrant™ Cell-Labeling Solutions (Molecular Probes, Eugene, OR, USA) to track both stem cells (*red*, DiD) and a second cell type (*green*, DiO). Cells are suspended at a concentration of 1×10^6/mL in serum-free media with 5 μL cell-labeling solution/mL

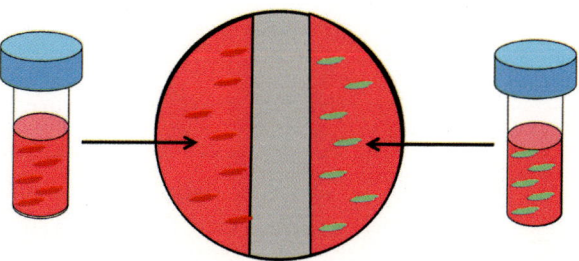

Fig. 3 Seeding of Cell Types A and B into a 3.2 cm^2 tissue culture plastic well with hydrogel divider in place

3.3 Cell Seeding

All steps should be performed in the dark, preferably.

1. For hydrogel divider samples:
   (a) Pipette 0.5 mL of Cell Type A cell suspension (5×10^4 total cells) directly onto the tissue culture plastic surface on the left side of the well, to the left of the divider (Fig. 3).
   (b) Pipette 0.5 mL of Cell Type B cell suspension (5×10^4 total cells) onto the tissue culture plastic surface on the right side of the well (Fig. 3).
   (c) Incubate for 30 min to allow for cell settling and attachment.
   (d) Slowly add enough F/S culture media to completely cover the hydrogel divider.

(e) To allow for temporally controlled cell–cell contact, the hydrogel divider can later be removed by sliding a scalpel blade along the circumference of the well, severing the gel from the plastic well. Discard the hydrogel.

2. For analysis of the behavior of individual subpopulations following culture:

(a) Pipette a 10 μL droplet of Cell Type A cell suspension onto a Thermanox coverslip.

(b) Pipette a 10 μL droplet of Cell Type B cell suspension onto a second cover slip.

(c) Incubate cover slips at 37 °C for 30 min to allow cells to attach.

(d) Place the cover slip with Cell Type A on the right side of the well, to the right of the hydrogel divider.

(e) Place the cover slip with Cell Type B onto the left side of the well, to the left of the hydrogel divider.

(f) Add enough F/S culture media to the well to cover the hydrogel divider.

3. For monoculture control group samples:

(a) Pipette 1 mL of Cell Type A or Cell Type B (1×10^5 total cells) into each 3.8 cm^2 tissue culture plastic well.

(b) Add 1 mL F/S culture media to each well.

4. For direct co-culture control samples:

(a) In the dark, combine 50 % Cell Type A cell suspension and 50 % Cell Type B cell suspension in a sterile centrifuge tube.

(b) Pipette 1 mL of mixed suspension (1×10^5 total cells) into a 3.8 cm^2 tissue culture plastic well.

(c) Add 1 mL F/S media to each well.

4   Notes

1. PBS should be sterile filtered on the same day it is going to be used.

2. Antibiotics used should include, at minimum, penicillin–streptomycin. The addition of other antibiotics is optional.

3. Culture media should be stored at 4 °C when not in use. It can be stored for up to 1 month prior to use.

4. Vybrant™ Cell-Labeling Solutions should be stored at 4 °C upon arrival and should be stored in the dark.

5. Filter PBS using a sterile filter with 0.22 μm pores.

6. Sterile-filtered PBS can be stored at room temperature in a glass container following filtering for up to 6 months.

7. Allow agarose solution to cool/gel for 10–20 min at room temperature. Be sure to examine gel every 5 min, and begin step 5 as soon as the gel is formed, in order to avoid gel dehydration.

8. For co-culture, the primary cells used are typically between passage 2 and passage 5.

9. Tilt cell pellet upwards and aspirate slowly from the bottom side of the conical tube to avoid disturbing the pellet.

10. For accuracy, count cells at least three times, and take the average cell number over the three counts.

11. Molecular Probes has published a short list of optimized cell staining incubation times for specific cell lines that are less than 20 min. Check this list before proceeding with a 20-min incubation period.

12. Start by incubating cells in labeling solution for 20 min and subsequently optimize if uniform labeling is not achieved after 20 min of incubation.

13. Be sure to resuspend cells in solution periodically to prevent cell clumping.

Acknowledgements

This work was supported by NIH grant R01-AR055280 (H.H.L.) and the NSF GRFP (D.R.B.).

References

1. Bhatia SN, Balis UJ, Yarmush ML, Toner M (1999) Effect of cell–cell interactions in preservation of cellular phenotype: cocultivation of hepatocytes and nonparenchymal cells. FASEB J 13:1883–1900

2. Lu HH, Jiang J (2006) Interface tissue engineering and the formulation of multiple-tissue systems. Adv Biochem Eng Biotechnol 102:91–111

3. Prockop DJ (1997) Marrow stromal cells as stem cells for nonhematopoietic tissues. Science 276:71–74

4. Pittenger MF, Mackay AM, Beck SC, Jaiswal RK, Douglas R, Mosca JD, Moorman MA, Simonetti DW, Craig S, Marshak DR (1999) Multilineage potential of adult human mesenchymal stem cells. Science 284:143–147

5. Caplan AI, Dennis JE (2006) Mesenchymal stem cells as trophic mediators. J Cell Biochem 98:1076–1084

6. Fan H, Liu H, Toh SL, Goh JC (2008) Enhanced differentiation of mesenchymal stem cells co-cultured with ligament fibroblasts on gelatin/silk fibroin hybrid scaffold. Biomaterials 29:1017–1027

7. Richardson SM, Walker RV, Parker S, Rhodes NP, Hunt JA, Freemont AJ, Hoyland JA (2006) Intervertebral disc cell-mediated mesenchymal stem cell differentiation. Stem Cells 24:707–716

8. Plotnikov EY, Khryapenkova TG, Vasileva AK, Marey MV, Galkina SI, Isaev NK, Sheval EV, Polyakov VY, Sukhikh GT, Zorov DB (2008) Cell-to-cell cross-talk between mesenchymal stem cells and cardiomyocytes in co-culture. J Cell Mol Med 12:1622–1631

9. Sheng H et al (2008) A critical role of IFN-gamma in priming MSC-mediated suppression of T cell proliferation through up-regulation of B7-H1. Cell Res 18:846–857

10. Csaki C, Matis U, Mobasheri A, Shakibaei M (2009) Co-culture of canine mesenchymal stem cells with primary bone-derived

osteoblasts promotes osteogenic differentiation. Histochem Cell Biol 131:251–266

11. Aguirre A, Planell JA, Engel E (2010) Dynamics of bone marrow-derived endothelial progenitor cell/mesenchymal stem cell interaction in co-culture and its implications in angiogenesis. Biochem Biophys Res Commun 400:284–291

12. Proffen BL, Haslauer CM, Harris CE, Murray MM (2013) Mesenchymal stem cells from the retropatellar fat pad and peripheral blood stimulate ACL fibroblast migration, proliferation, and collagen gene expression. Connect Tissue Res 54:14–21

13. Wang T, Xu Z, Jiang W, Ma A (2006) Cell-to-cell contact induces mesenchymal stem cell to differentiate into cardiomyocyte and smooth muscle cell. Int J Cardiol 109:74–81

14. Zhang Y, Chai C, Jiang XS, Teoh SH, Leong KW (2006) Co-culture of umbilical cord blood CD34+ cells with human mesenchymal stem cells. Tissue Eng 12:2161–2170

15. Lu HH, Wang IE (2007) Multi-scale co-culture models for orthopaedic interface tissue engineering. In: Gonsalves KE et al (eds) Biomedical nanostructures. Wiley, New York

16. Lee IC, Wang JH, Lee YT, Young TH (2007) The differentiation of mesenchymal stem cells by mechanical stress or/and co-culture system. Biochem Biophys Res Commun 352:147–152

17. Zhang L, Tran N, Chen HQ, Kahn CJ, Marchal S, Groubatch F, Wang X (2008) Time-related changes in expression of collagen types I and III and of tenascin-C in rat bone mesenchymal stem cells under co-culture with ligament fibroblasts or uniaxial stretching. Cell Tissue Res 332:101–109

18. Jiang J, Nicoll SB, Lu HH (2005) Co-culture of osteoblasts and chondrocytes modulates cellular differentiation in vitro. Biochem Biophys Res Commun 338:762–770

19. Wang IE, Lu HH (2006) Role of cell-cell interactions on the regeneration of soft tissue-to-bone interface. Conf Proc IEEE Eng Med Biol Soc 1:783–786

20. Wang IE, Shan J, Choi R, Oh S, Kepler CK, Chen FH, Lu HH (2007) Role of osteoblast-fibroblast interactions in the formation of the ligament-to-bone interface. J Orthop Res 25:1609–1620

21. Hamilton SK, Bloodworth NC, Massad CS, Hammoudi TM, Suri S, Yang PJ, Lu H, Temenoff JS (2013) Development of 3D hydrogel culture systems with on-demand cell separation. Biotechnol J 8:485–495

22. Bhatia SN, Yarmush ML, Toner M (1997) Controlling cell interactions by micropatterning in co-cultures: hepatocytes and 3T3 fibroblasts. J Biomed Mater Res 34:189–199

23. Takayama S, McDonald JC, Ostuni E, Liang MN, Kenis PJ, Ismagilov RF, Whitesides GM (1999) Patterning cells and their environments using multiple laminar fluid flows in capillary networks. Proc Natl Acad Sci U S A 96:5545–5548

24. Ma SH, Lepak LA, Hussain RJ, Shain W, Shuler ML (2005) An endothelial and astrocyte co-culture model of the blood-brain barrier utilizing an ultra-thin, nanofabricated silicon nitride membrane. Lab Chip 5:74–85

25. Kane BJ, Zinner MJ, Yarmush ML, Toner M (2006) Liver-specific functional studies in a microfluidic array of primary mammalian hepatocytes. Anal Chem 78:4291–4298

26. Bogdanowicz DR, Lu HH (2013) Studying cell–cell communication in co-culture. Biotechnol J 8:395–396

Methods in Molecular Biology (2014) 1202: 37–55
DOI 10.1007/7651_2014_76
© Springer Science+Business Media New York 2014
Published online: 1 April 2014

Multiwell Plate Tools for Controlling Cellular Alignment with Grooved Topography

Camila Londono, John Soleas, Petra B. Lücker, Suthamathy Sathananthan, J. Stewart Aitchison, and Alison P. McGuigan

Abstract

In many tissues, cells must be aligned for proper function. This alignment can occur at the cellular and/or subcellular (protein/molecular) level. The alignment of cytoskeletal components, in fact, precedes whole cell alignment. A variety of methods exist to manipulate cytoskeletal and whole cell alignment; one of the simplest and most predictable involves seeding adherent cells onto defined substrate topography. We present here two methods to create grooved multiwell plates: one involving microfabrication, which allows for custom design of substrate topography, and a simpler, inexpensive method using commercially available diffraction gratings. We also include methods for manual and automatic quantification of cell alignment.

Keywords: Cell alignment, Substrate topography, High-throughput, Actin, Microtubules, Microfabrication, Microgrooves

1 Introduction

Tissue function often depends on the concerted behavior of individual cells. Though many different mechanisms exist to harmonize cellular behavior, cellular alignment is one basic way in which the actions of separate cells can be coordinated. One example is the stromal keratocytes in the cornea, which are responsible for the secretion of aligned collagen fibrils, critical for corneal transparency and strength. In vitro studies have shown that the keratocytes must be prealigned for the fibrils to align and be packed correctly (1). Many other tissues, including myocardium (2), muscles (3), kidney (4), and blood vessels (5), contain components with aligned and precisely arranged cells. Replicating this alignment, therefore, is a necessary part of recreating each tissue's functionality in vitro.

Cellular alignment can be measured at a number of scales. The most common and obvious metric, at the cellular scale, involves looking at whole cell bodies, which become elongated and align along their longest axis (Fig. 1a, b. ARPE-19 brightfield image on flat vs. grooved). In the absence of whole cell elongation, alignment

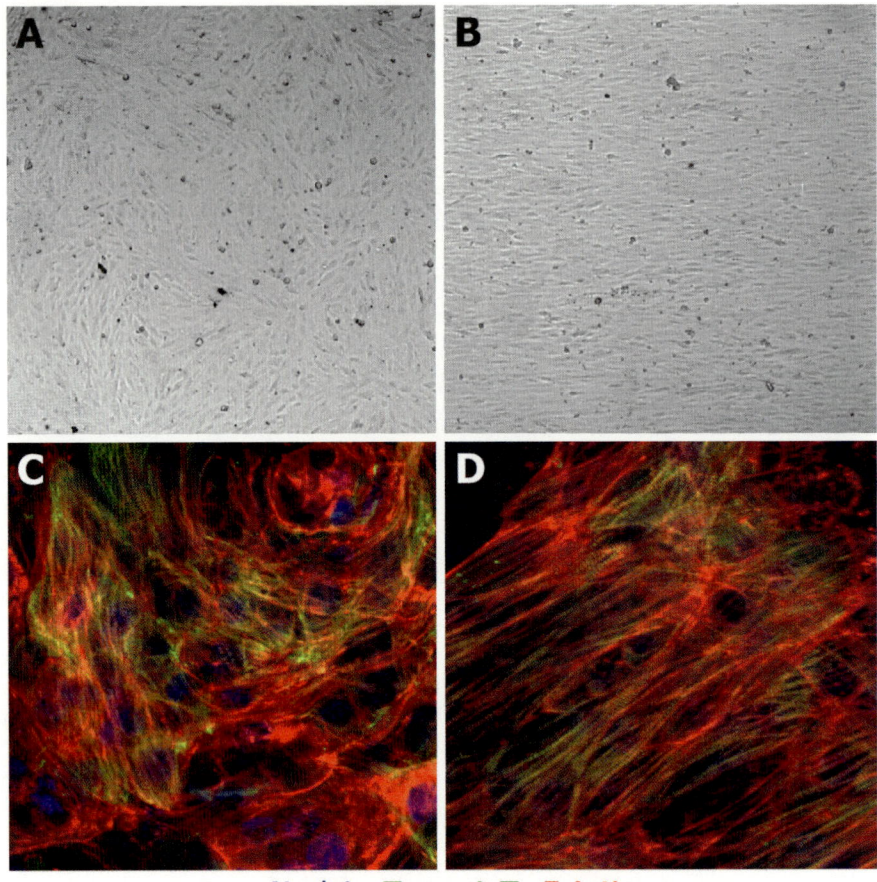

Nuclei   TroponinT   F-Actin

Fig. 1 Substrate topography can be used to produce alignment of both whole cells and the cytoskeleton: (**a**, **b**) Human retinal pigment epithelial cell line ARPE-19 seeded on a flat substrate (**a**), display random organization, while those seeded on a grooved substrate (**b**) become aligned to the direction of the grooves. (**c**, **d**) Cardiomyocyte actin stress fibers are disorganized when cells are seeded on flat substrates (**c**), while they align to the direction of the grooves when the cells are seeded on grooves substrates (**d**)

of polarized molecular components, such as cytoskeletal components or the planar cell polarity complexes, may occur at the subcellular scale (6). Indeed, whole cell alignment is thought to arise as a consequence of the subcellular alignment of the cytoskeleton and focal adhesions. Cytoskeletal alignment most often refers to stress fibers composed of actomyosin bundles associated with cell-substrate adhesion complexes. Stress fibers are known to be responsible for contractile forces in cells (7). The alignment of these stress fibers (in the direction of the grooves on grooved topography (Fig. 1c, d—cardiomyocyte triple staining on flat and grooved substrates)) is often more obvious than that of whole cells (8). A less examined but no less important measure of cytoskeletal alignment is microtubule alignment. In fibroblasts, microtubules become aligned on the groove regions (as opposed to ridges) on grooved substrates. Their alignment precedes actin fiber and whole cell alignment (9). These measures of cell alignment are often observed in combination with one another, and in most cases, act jointly to affect cellular behavior.

Cell and cytoskeletal alignment have been manipulated in vitro in a number of ways. Shear stress, for example, has been used to align endothelial cells (10) and is believed to refine cellular alignment in the airway after birth (6). Seeding cells on deformable substrates and then subjecting those to cyclic stretching also results in aligned endothelial cells (11). Electrical fields have been shown to play a role in the alignment of lens and corneal epithelial cells (12). Still, the easiest and most versatile method for manipulating cell alignment is seeding cells on controlled topographic features. Substrate topography can be used to create both simple and complex cell organizations. A number of studies have demonstrated that grooved topography results in cellular and cytoskeletal alignment (1, 10, 12–14) in a variety of cell types. Here, we present two variations on a method to create multiwell plates with grooved topography for high-throughput applications. The first allows for custom-designed patterns using photolithography, while the second makes use of commercially available, inexpensive diffraction grating sheets to obviate costly microfabrication. We also describe metrics that can be used for characterization of cellular alignment and summarize currently available software for analysis of cellular alignment.

2 Materials

2.1 Large Equipment

1. Sonicator.
2. Oven.
3. Spin coater.
4. Hot plate.
5. Mask aligner (e.g. Suss MicroTec MA6 Mask Aligner).
6. Inductively coupled plasma reactive ion etcher (e.g. Trion Phantom II RIE/ICP system).
7. Desiccator.
8. Microscope (brightfield/transmitted light, fluorescence, confocal, high-throughput depending on application).
9. Plasma cleaner.
10. UV sterilization cabinet, or alternatively, biosafety cabinet equipped with UV light.

2.2 Substrate Fabrication/ Preparation

1. Chrome-glass photomask (for custom feature design).
2. Silicon wafer(s).
3. Acetone.

4. Isopropyl alcohol.

5. Deionized water.

6. Nitrogen gas.

7. Hexamethyldisilazane (MicroPrime MP-P20 HMDS, Shin-etsu, Japan).

8. MICROPOSIT® S1811® positive photoresist (Shipley, PA).

9. MICROPOSIT® MF®-321 Developer (Shipley, PA).

10. Holographic diffraction grating film (Edmund Optics, Catalog No. #40-267).

11. Polycarbonate sheet, approximately 6 mm thick, or glass, trimmed to fit into desiccator and oven.

12. Transparent packing tape.

13. Laboratory labelling tape, 25.4 mm wide.

14. Scotch tape.

15. Glass pasteur pipette.

16. (Tridecafluoro-1,2,2-tetrahydrooctyl)-1-tricholorosilane (United Chemical Technologies, PA, Catalog No. T2492-KG).

17. Sylgard® 184 Silicone Elastomer kit (polydimethylsiloxane, PDMS) (Dow Corning, MI, catalog number depends on distributor).

18. Two flat spatulas.

19. Scissors.

20. Scalpel.

21. Cutting board.

22. Disposable syringe, 20 mL (e.g. BD 302830).

2.3 Plate Assembly

1. Bottomless 96-well plates (Greiner Bio One, Catalog No. 655000).

2. 96-well plate polystyrene lids (Corning, Catalog No. 3931).

3. Hexane (not Hexanes or mixture of isomers) (Sigma-Aldrich, Catalog No. 208752).

4. Disposable syringe, 5 mL (e.g. BD 309646).

5. Plastic transfer pipette (e.g. VWR 16001–172 or 16001–176).

6. 15 mL conical centrifuge tube (e.g. Falcon 352196).

7. Glass pipette.

8. Aluminum foil.

9. Rack portions of two P200 tip box (e.g. Fisherbrand Redi-tip pipet box).

10. Printout of alignment guidance drawing (Fig. 2a).

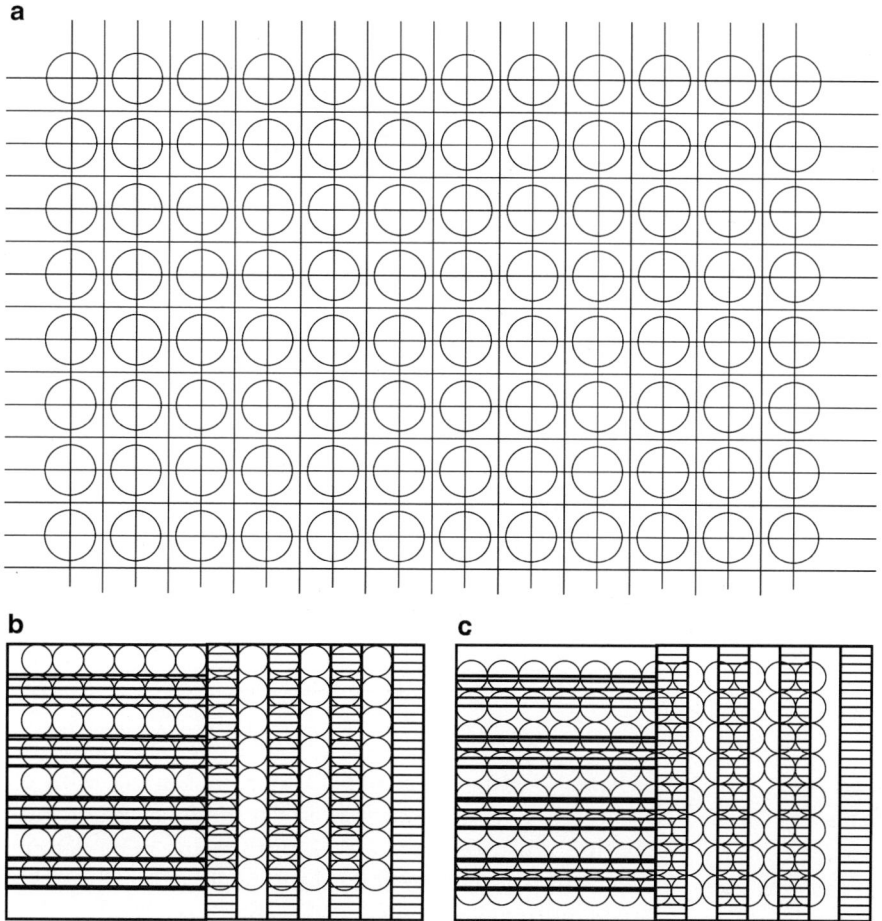

Fig. 2 Feature alignment tools and possible configurations: (**a**) Plate alignment guide: Light diffraction produced by the grooves can be used to align the PDMS substrates and the *bottomless plates* by placing the sheet on the alignment guide and visually aligning the diffractive regions within *straight lines*, and then aligning the wells in the *bottomless plate* to the *circles*. (**b**, **c**) Possible plate configurations: A mask that alternates between grooved and flat regions can be used to create wells that are either fully grooved or fully flat (**b**), or half-grooved, half-flat (**c**)

11. Kimwipes.

12. Polycarbonate sheet, approximately 2–3 mm thick, 114 mm long × 77 mm wide, one per Plate.

13. 4″ or 4.5″ Bar clamps, four per Plate.

2.4 Cell Culture

1. 70 % ethanol.

2. Phosphate buffered saline (PBS).

3. 10 μg/mL stabilized bovine fibronectin diluted in PBS (Biomedical Technologies Inc., Catalog No. BT-226S).

4. Trypsin or Trypsin-EDTA (as appropriate for specific cell type).

5. Complete cell culture medium (as appropriate for specific cell type).

2.5 Cell Staining

1. Hoechst 33342 (Invitrogen Life Technologies, Catalog No. H3570).

2. Rhodamine phalloidin (Invitrogen Life Technologies, Catalog No. R415).

3. Anti-β-tubulin antibody (Abcam, Catalog No. ab7792).

4. Alexa488-conjugated anti-mouse antibody (Invitrogen Life Technologies, Catalog No. A11029).

5. Antibodies against cadherins or β-catenin, optimized for your cell types and samples.

6. CellMask orange plasma membrane stain (Invitrogen Life Technologies, Catalog No. C10045).

2.6 Software (All Optional Depending on Objectives)

1. L-edit layout editor for mask design (Tanner EDA, CA, USA).

2. AutoCAD (Autodesk, Inc.).

3. ImageJ ((15)).

4. Rose (Todd A. Thompson, Indiana University, http://mypage.iu.edu/~tthomps/).

3 Methods

3.1 Method 1: Silicon Master Microfabrication for Custom Features

1. Using L-edit or AutoCAD, design a mask with the desired features. Our mask contained flat sections and grooved sections with 2 μm wide lines with 2 μm spacing. The flat and grooved sections should be 9.02 mm wide to correspond with the spacing between wells in 96-well plates.

2. Print the design on chrome-glass to generate a photomask for features that are within this size range. Facilities like the Micro and Nanofabrication Facility at the University of Alberta can do this to order.

3. Sonicate a new silicon wafer in acetone for 3 min (*See* **Note 1**).

4. Rinse wafer in isopropyl alcohol for 1 min.

5. Rinse wafer in deionized water for 1 min.

6. Dry wafer using nitrogen gas.

7. Bake wafer at 105 °C for 2 min to dehydrate. Allow to cool.

8. Spin-coat adhesive primer hexamethyldisilazane (P20-HMDS) onto the wafer at 4,000 rpm for 40 s in order to improve adhesion of the photoresist to the wafer.

9. Spin-coat S1811 positive resist onto the wafer at 4,000 rpm for 40 s (Fig. 3a).

10. Soft-bake the silicon wafer at 105 °C on a hotplate for 2 min to remove solvents. Allow to cool.

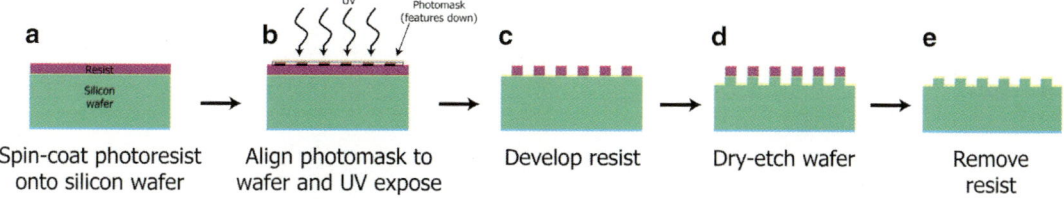

Fig. 3 Microfabrication of silicon masters (method 1): (**a**) A clean silicon wafer is spin-coated with photoresist. (**b**) The photomask is aligned with the silicon wafer and exposed to UV light. (**c**) The photoresist is developed, and nonexposed photoresist is washed away. (**d**) The silicon wafer is etched using a reactive ion etcher. (**e**) Exposed photoresist is washed away, leaving a silicon wafer with grooved features

11. Place photomask and resist-coated silicon wafer on the mask aligner. Expose the wafer to UV (365 nm wavelength) in hard contact mode for a selected exposure time which depends on the intensity of the UV lamp being used (for a UV lamp with 16.9 mW/cm² intensity, the exposure time was 10 s (*See* **Note 2**)) (Fig. 3b).

12. Remove the wafer from the mask aligner and develop in MF-321 developer while agitating gently for 40–50 s or until the resist is removed in all exposed areas (Fig. 3c).

13. Rinse the developed wafer in deionized water for 40–50 s. Inspect under an optical microscope to ensure all photoresist has been removed before proceeding.

14. Hard-bake the wafer 105 °C on a hotplate for 2 min to harden the resist. Allow to cool.

15. Etch the wafer using an inductively coupled plasma reactive ion etcher (RIE) at a pressure of 100 mTorr and RIE RF power of 120 W. Etchant flow rates should be the following: $SF_6 = 30$ sccm, $O_2 = 20$ sccm, $CHF_3 = 12$ sccm, He = 10 sccm. Etching time should be adjusted to obtain desired depth (30 s = ~180 nm deep, 75 s = ~600 nm deep, 300 s = ~3.13 µm deep in widest regions) (*See* **Note 3**) (Fig. 3d).

16. After achieving the desired depth, immerse etched master into acetone to remove remaining resist (Fig. 3e).

17. Rinse master in deionized water.

3.2 Method 2: Diffraction Grating Master Fabrication (Fig. 3)

1. Using scissors, cut two pieces of diffraction grating from the sheet (while still in its plastic packaging (*See* **Note 4**)). These pieces should be at least 84 mm long, and 116 mm wide. Also, their wide edge should come from the factory-cut edge of the sheet (red line in Fig. 4a, b) to ensure they are straight and can be aligned properly. Treat the diffraction grating sheet carefully to avoid damaging it as it is easily scratched.

2. Place a piece of transparent packing tape (approximately 140 mm long) glue side up on a solid surface.

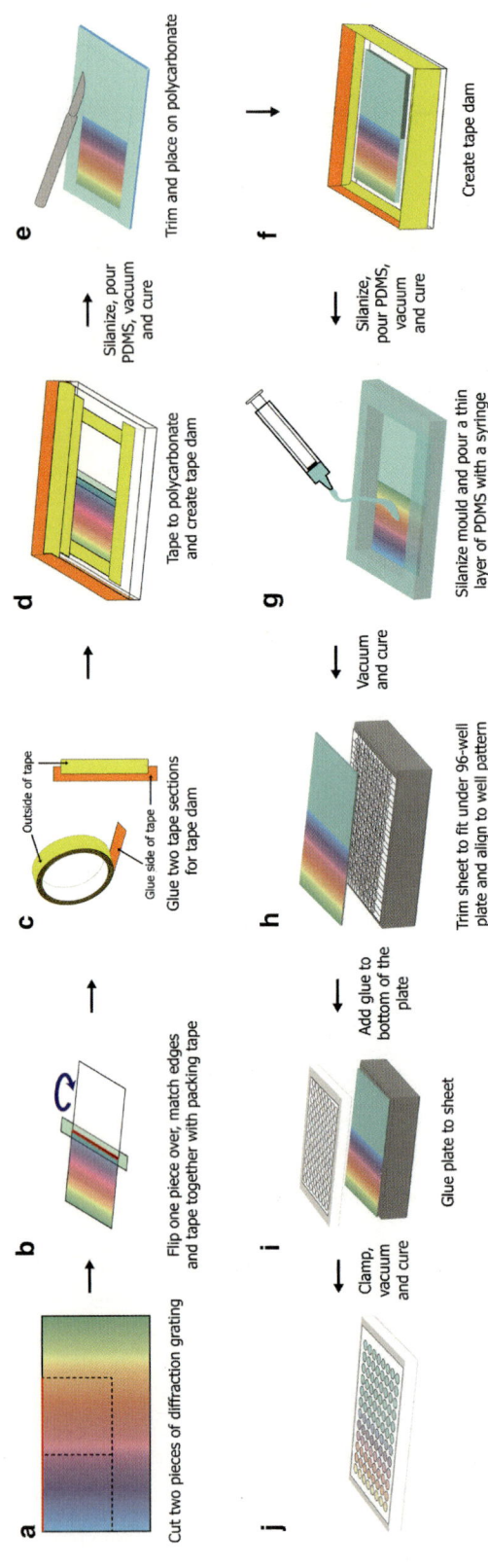

Fig. 4 Fabrication of diffraction grating masters (method 2) and plate assembly: (**a**–**f**) Diffraction grating master fabrication: A diffraction grating sheet (Edmund Optics) is trimmed (**a**), to produce two pieces; one of these is flipped, and they are taped to one another and to a support surface (**b**). Two pieces of laboratory labelling tape are taped to one another (**c**), to generate "walls," (**d**) for the support surface to constrain liquid PDMS. Once PDMS has been poured and cured, the resulting block must be trimmed (**e**), to the correct size, and used to create a mould (**f**). (**g**–**j**) Plate assembly: Using a silanized mould, PDMS should be extruded into a thin layer using a syringe (**g**). Once that sheet has cured, it should be trimmed to size and aligned to the plate alignment tool (Fig. 2a) (**h**). A bottomless plate should be glued onto the sheet (**i**), to generate a multiwell plate with grooved and flat topography (**j**)

3. If necessary, check the directions/location of the grooves on the diffraction grating pieces by very lightly spraying the sheet with alcohol and gently shaking the sheet in the air, while grabbing it from one corner to prevent damaging the grooves. Capillary action will cause the droplets to move into the grooves in the grooved section, making the alcohol droplets have straight rather than rounded edges and making the grooves visible (*See* **Note 5**). Allow to dry.

4. Place the straight edge (i.e. the side that was not cut with scissors—red line in Fig. 4a, b) of one of the diffraction grating pieces (with grooves facing up) on the tape such that it lies lengthwise approximately along the center of the packing tape (Fig. 4b). Slowly, lower the rest of the sheet. If necessary, press lightly on the corners of the sheet and along the back to the tape to ensure it sticks. It is acceptable for a few bubbles to remain between the tape and the diffraction grating.

5. Flip the remaining piece of diffraction grating over (this piece should now have grooves facing down). Align the straight edge very carefully to make contact with the edge of the piece already stuck on the packing tape, ensuring they do not overlap. Lower the second piece onto the tape as previously. This generates a surface where half of the sheet is grooved and half is flat (Fig. 4b).

6. Fold over the ends of the packing tape onto itself to cover all the exposed glue and make the master sheet easier to handle.

7. Holding the resultant master sheet by the packing tape to prevent damaging it, place it on a piece of polycarbonate or glass for support.

8. Using laboratory labelling tape, tape the master sheet to the support material. Between 2 and 5 mm of the edges of the master sheet should be covered with tape. Ensure that the sheet is not bent, and that no air pockets remain between the support and the sheet.

9. Cut two pieces of laboratory labelling tape to be slightly longer than the support material (i.e. polycarbonate or glass piece) (Fig. 4c). Place one sticky side up on a solid surface, and then stick the other one, face down, at about the midline of the first piece (*See* **Note 6**). The resultant piece should be about 37 mm wide, and have two sticky regions facing in opposite directions. Stick one of those regions on the top face of the support piece along the edge. Bend up to create the first wall of the tape dam (Fig. 4d).

10. Repeat on all sides, overlapping at the corners. You may need to pinch the corners to tighten the tape dam.

11. Add tape (preferably all one piece) to the outside of the walls to create a periphery and prevent the walls from collapsing outwards.

12. If needed, add tape to the walls or inner corners. Ensure no glue remains uncovered towards the inside of the dam (where the grating is located) to prevent PDMS from not curing (*See* **Note 7**).

13. Place the master and support structure into a desiccator in a fume hood.

14. In a fume hood, use a glass pasteur pipette to add two or three drops of (tridecafluoro-1,2,2-tetrahydrooctyl)-1-tricholorosilane to a scintillation vial. Place this vial and the pasteur pipette in the desiccator, outside of the master and tape dam (*See* **Note 8**).

15. Apply vacuum to the desiccator for 5–10 min, then close valve and allow to remain in vacuum for 3 h to silanize the master (*See* **Note 9**). This generates a nonstick surface for subsequent moulding of PDMS (*See* **Note 9**).

3.3 Mould Fabrication

1. Prepare PDMS by mixing the Sylgard®184 elastomer and curing agent at a mass ratio of 10:1. There should be enough PDMS to create a piece about 4–5 mm deep over the master. The necessary amount will depend on the size of the containers, but 60 g is appropriate for 15 cm dishes (for the 4″ silicon wafers).

2. Degas PDMS under vacuum until all bubbles have disappeared. Watch during degassing to prevent spillage. If foam rises too high during degassing, allow some air into vacuum chamber, and then resume pulling vacuum.

3. Once the PDMS is fully degassed, pour over master. If using a silicon master, the wafer should be placed features side up on a 15 cm dish. Since this master will not be stuck to the container, pour PDMS in the center first to add some weight to the master and prevent too much PDMS from seeping underneath.

4. Degas PDMS once again after pouring. Make sure all bubbles have been removed to ensure full replication of the features.

5. Cure PDMS at 60 °C for at least 3 h or overnight.

6. Once the PDMS has become solid, remove from the oven.

7. If the master is in a tape dam, remove tape and peel PDMS in the direction of the grooves to obtain a surface with a negative imprint of the master. If it is in a dish, use the sharp end of a spatula to separate the mould from the edges and lift from the dish (*See* **Note 10**). A second spatula may be used to help in removing the PDMS and master from the dish. Flip the dish over, and lightly bend it to encourage the PDMS to detach

from the dish. Once removed, trim the edges enough to ensure a flat surface.

8. *For diffraction grating moulds only:* Using a scalpel, trim the PDMS negative to the outer edge of the grooved/flat regions, which are surrounded by tape marks (Fig. 4e). Place on polycarbonate or glass support, and create another tape dam as in Fig. 3d (Fig. 4f) (*See* **Note 11**). Repeat steps 1–5 in Section 3.3 to create a deep mould from your negative. The depth will be the same as that of your negative. Use a spatula to press along the edges of the PDMS negative. This will allow you to peel the negative from the mould.

9. *For silicon wafer moulds only:* Moulds made with a silicon master may have a thin layer of PDMS on the bottom side of the wafer. Carefully, cut this thin layer from around the wafer using a scalpel, and remove the layer (*See* **Note 12**). Bend the mould very lightly to remove the silicon wafer. Place the mould and the master, both feature side up, in separate dishes.

10. Silanize the newly made moulded PDMS as per above (Steps 14–15 in Section 3.2), by placing in a desiccator with two or three drops of (tridecafluoro-1,2,2-tetrahydrooctyl)-1-trichlorosilane in a scintillation vial and leaving under vacuum for 3 h. This prevents uncured PDMS in further steps from sticking to the cured PDMS mould.

3.4 Substrate Fabrication and Plate Assembly

1. Mix about 20–25 g of PDMS (depending on the size of your mould) at a ratio of 10:1 (w/w) elastomer:curing agent. Degas under vacuum until all bubbles are gone.

2. Using a 20 mL syringe, extrude PDMS to form a thin layer over the mould to generate the sheet that will be the bottom of your plate (Fig. 4g). Move all over the mould as you press on the plunger to cover the whole surface with PDMS without leaving holes.

3. Degas once again and ensure there are no holes in the layer left after degassing. Add extra PDMS and degas further if holes are present. Cure at 60 °C for at 3 h or overnight.

4. Once the PDMS has solidified, use a spatula to lift the edges of the thin layer from the mould. Peel layer.

5. Using a scalpel, trim the layer to fit under the plate. There are protruding notches along the inside of the edge of the plates. Ensure PDMS is cut small enough to fit inside those notches to prevent leaking (*See* **Note 13**).

6. Use a piece of scotch tape (or other low-adhesive type) to create a loop (glue side facing out) that can approximately fit around all your fingers. Remove any dust from PDMS sheet by lightly dabbing the sheet with the tape loop. The tape can be detached by pulling or rolling it while making sure to not touch the sheet

with your fingers. Do not press the tape down as this will potentially damage the surface grooves. Keep track of grooved locations, though light diffraction should make them obvious.

7. Cover both of the pipette tip racks in foil, making sure the top is flat and wrinkle-free.

8. Print out the alignment guidance drawing (Fig. 2a) and tape to one of the foil-covered racks.

9. Place the PDMS sheet on the printout and align features to the wells (Fig. 2b, c—plate configurations and Fig. 3h) using light diffraction to distinguish between grooved and flat regions.

10. Prepare a small amount of PDMS as above (10:1 w/w ratio).

11. Within a fumehood, draw liquid PDMS into a 3 mL syringe, and using the syringe, add 0.3 mL of PDMS into a 15 mL centrifuge tube (*See* **Note 14**).

12. Using a glass pipette, add 2.7 mL of hexane to the centrifuge tube for a mixture that is nine parts hexane to one part PDMS (v/v). Close very tightly and shake to mix. Ensure that PDMS and hexane are combined properly and no unmixed PDMS is left at the bottom or sides of the tube. Ready bottomless plate and second foil-covered pipette rack in fumehood.

13. Use the plastic bulb transfer pipette to suction the hexane: PDMS adhesive mixture, and then smear it with the side of the tube part of the pipette on the foil covering the second pipette tip rack. Approximately 1 mL of mixture should be enough to cover the foil.

14. Quickly, without allowing the liquid to dry, rub the bottom of the bottomless plate over the foil to transfer the hexane:PDMS adhesive from the foil to the plate. Check the entire bottom of the plate has been wetted; if not, carefully add more of the adhesive where necessary using the plastic transfer pipette.

15. Allow the plate to dry, with the wetted surface facing down, for about 15 s. The plate can now be removed from the fumehood if necessary.

16. Using the alignment guide, place the plate on the PDMS sheet. Press it down firmly, making sure not to move it horizontally, which would damage the features (Fig. 3i). Visually check that the sheet has stuck down (there should be no diffraction pattern visible between wells), and press lightly on any sections that are not stuck.

17. Place under vacuum for 5–10 min.

18. After turning the plate upside down, place a kimwipe over the PDMS sheet, and then the 114 mm × 77 mm polycarbonate sheet over the kimwipe. Make sure the polycarbonate sheet is within the notches on the plate by folding kimwipe to check.

Use the clamps (one on each end, as close to the center as possible, and two on the line between grooved and flat) to apply pressure to the polycarbonate and plate (*See* **Note 15**).

19. While still clamped, cure at 60 °C overnight. The plate will be ready to use after curing (Fig. 3j).

3.5 Cell Seeding

1. Spray 70 % ethanol in the plate and on a polystyrene lid and allow to rest for 20 min.

2. Remove ethanol by shaking on to a paper towel or a sink.

3. *Optional*: Place uncovered plate in a plasma cleaner and plasma clean for 10 min (*See* **Note 16**).

4. UV sterilize the uncovered plate and the lid for 10 min.

5. While plate is being sterilized, prepare a 10 µg/mL stabilized fibronectin solution in PBS. Once sterilization is complete, move plate to a biosafety cabinet and add 200 µL of fibronectin solution to each well. Allow to rest in an incubator at 37 °C and 5 % CO_2 for 1 h.

6. Trypsinize cells with enough time to ensure they will be ready to seed once the hour has passed. Resuspend cells in their culture medium so they will create a confluent sheet with 100 µL of cell suspension. For ARPE-19 cells, this is between 180,000 and 210,000 cells/mL. For HUVEC, 200,000 cells/mL is sufficient. Adjust according to cell type.

7. Remove fibronectin, wash once with PBS, and add 100 µL of cell suspension to each well. Place in the incubator for 1 h.

8. Add 100 µL of complete cell culture media to each well.

9. Cell alignment is usually observed soon after cell attachment but this may vary with cell type.

3.6 Manual Alignment Analysis

Before quantifying alignment of whole cells or actin cytoskeleton, high-quality images of greater magnification than 10× must be taken. To approximate the sample faithfully, we suggest at minimum 20 images be randomly taken of each individual sample. Out of these 20, ten should be randomly selected and placed in a different computer folder in a blinded fashion (that is to say, by a party removed from the microscopy) and then quantified as follows:

3.6.1 For Whole Cell Alignment

1. Open ImageJ (National Institute of Health).

2. Under the *File* subheading select open, and choose the first image from the referred folder.

3. Under the *Analyze* subheading, select *Set Measurements*.

    (a) A popup menu will appear.

4. Toggle *Fit ellipse*.

5. Click OK at the bottom.

6. Under the Subheading bar there is a row of buttons

   (a) Click the fourth button from the right, *Freehand selections*.

7. On your opened image, trace the cell outline (the visible cell membrane) as carefully as possible for a cell in the top most right corner.

   (a) A yellow line will mark where your cursor has been.

8. Click Ctrl + M on your keyboard.

9. A second window will open displaying the angle measurement.

10. Repeat steps 7 and 8 for a cell in the top left, bottom right, bottom left, and middle of the image.

11. After 5 separate cells have been outlined and their angles quantified, open the *File* subheading and click *Open Next*.

12. This should open the next image.

13. Repeat steps 7–11.

14. The measurements should be added to the same pop-up window.

15. After all ten images have been processed in this manner, click Ctrl + A in the measurement popup window to select all the data. Copy this data into an Excel spreadsheet, making sure to label each column appropriately.

16. Excel column charts can be used to graph the data. However, angular data is more commonly represented with rose diagrams. Software such as Rose (http://pages.iu.edu/~tthomps/programs) can be used to create those.

17. Repeat process as necessary.

3.6.2 For Cytoskeletal Alignment

Images should be fluorescent micrographs stained for F-Actin (phalloidin) or microtubules (β-tubulin) (*See* **Note 17**).

1. Repeat steps 1–6 from the whole cell alignment section.

2. On your opened image, trace an individual actin stress fiber or microtubule as carefully as possible (a good laser mouse or pen tablet makes a difference) of a cell located in the top most right corner of the image.

   (a) A yellow line will mark where your cursor has been.

3. Perform steps 8–9 from above.

4. Trace 5 fibers per cell in 5 cells per image (top right, top left, bottom right, bottom left, and center cells is ideal for consistency from image to image).

5. Perform steps 11–17 from above.

3.7 Automated Analysis of Whole Cell Alignment

Cell membrane staining may be used to automatically detect the shape and major axis direction of all objects in an image. Automated analysis requires extremely high-quality confocal images in which cell boundaries are sharply defined. Discontinuities or high background make it difficult to automatically define cells. Lower-quality images will affect the accuracy of or even prevent measurements (*See* **Note 18**).

1. Open ImageJ (National Institute of Health).

2. Under the *File* subheading select open, and load your selected image.

3. Press Ctrl + Shift + T, or go to the *Image* subheading, then *Adjust*, then select *Threshold*

    (a) The threshold window will appear.

    (b) Check the *Dark background* checkbox.

4. Adjust the threshold values to highlight all the cell membranes

    (a) Click *Apply*.

5. Under the *Process* subheading, select *Binary*, and then *Make Binary*.

6. Under the *Process* subheading, select *Binary*, and then select *Skeletonize*.

7. Click on the *Analyze* menu, and then on *Set Measurements*

    (a) Toggle *Fit Ellipse*.

8. Click OK at the bottom.

9. Go to the *Analyze* subheading, then click on *Analyze particles*.

    (a) Make sure that *Include holes* and *Exclude on Edges* are not checked.

    (b) Change the zero in the *Size (pixel^2)* text box to exclude small debris (a number from 6 to 10 is a good starting point, but this needs to be optimized to the images being analyzed).

10. Click *OK*.

11. A second window will open displaying the angle measurements for all the cells in the image.

12. Repeat steps 15–17 from Section 3.6.1.

3.8 Alternative Analysis Methods

Though we describe above how to quantify cellular and cytoskeletal alignment using ImageJ, other computer software can be used as well. We summarize some of the available software options in Table 1. Software that is not free is often provided with microscopes or high content screening tools by each manufacturer, and may be able to perform cytoskeletal analysis as well as whole cell analysis.

Table 1
Software capable of analyzing cell and cytoskeletal alignment

Software	Created by	Link	Free?
ImageJ	National Institutes of Health	http://rsbweb.nih.gov/ij/	Yes
CellProfiler	Broad Institute	http://www.cellprofiler.org/	Yes
Packing analyzer	Benoit Aigouy (Max Planck Institute of Molecular Cell Biology and Genetics)	Email: benoit.aigouy@univ-amu.fr for a copy	Yes
SIESTA	Jennifer Zallen (Memorial Sloan Kettering Cancer Center) and Rodrigo Fernandez-Gonzalez (University of Toronto)	Email: rodrigo.fernandez.gonzalez@utoronto.ca for a copy	Yes
MetaMorph®/ MetaXpress®	Molecular Devices, LLC	http://www.moleculardevices.com/Products/Software/Meta-Imaging-Series/MetaMorph.html	No
Cellomics® cell morphology BioApplication	Thermo Fisher Scientific Inc.	http://www.thermoscientific.de/com/cda/product/detail/1,,10141755,00.html	No
Acapella® high content imaging and analysis software	PerkinElmer Inc.	http://www.perkinelmer.ca/en-ca/pages/020/cellularimaging/products/acapella.xhtml	No
IN cell analyzer or IN cell investigator	GE Healthcare Bio-Sciences Corp.	http://www.gelifesciences.com/webapp/wcs/stores/servlet/catalog/en/GELifeSciences-ca/brands/in-cell-analyzer	No
MATLAB® (image processing toolbox—regionprops)	The MathWorks, Inc.	http://www.mathworks.com/products/matlab/	No

4   Notes

We have identified the following issues that can arise during plate fabrication and use:

1. The wafer cleaning protocol described here is only suggested for use with new silicon wafers. If a recycled wafer is used instead, a thorough and comprehensive cleaning with Piranha solution or RCA-based cleaning will be required to remove any contaminants from the wafer surface. Protocols for cleaning

recycled wafers are usually available at microfabrication facilities and may be adapted and optimized for specific needs.

2. The exposure time can be calculated by dividing dose by intensity. The dose is specific to the selected photoresist and is usually found from the resist data sheet. It depends on the thickness of resist that has been spin-coated. The intensity is the UV lamp's intensity.

3. Etchant flow rates for the reactive ion etcher (step 15 in Section 3.1) may need to be optimized to obtain the desired etching rate.

4. The diffraction grating sheet is quite fragile, and as such, it must be treated carefully to prevent damaging the grooves. Both pressure and oil from fingerprints could result in losing the grooves. Check for damages by observing the sheet. Undamaged regions will appear to be rainbow-colored due to light diffraction. Damaged regions will have no rainbow.

5. Though alcohol can be used to determine the direction of the grooves (as per step 3 in Section 3.2), this should only be done if strictly necessary. The alcohol should be prepared with deionized water to prevent residue being left behind after the alcohol dries.

6. In step 9 in Section 3.2, a second person may be useful to hold the first piece of tape before adding the second one. Alternatively, cut the tape to be slightly longer than necessary, fold the ends over, and attach them to the solid surface to prevent movement.

7. PDMS that makes contact with the glue from the tape in the tape dam will not cure. It is important that as much of the glue surface be covered as possible, and also that the tape dam and support are big enough so that parts covered with uncured PDMS can be cut off. Also, using the wrong kind of tape can affect curing of the whole slab, so be sure to use labelling tape only.

8. Silanizing is important to prevent PDMS from sticking non-reversibly to the masters and moulds. Care should be taken to not add more than three drops of silane to prevent excess deposition. Although the objects to be silanized may be left under vacuum for longer than the suggested 3 h, it is possible that silane deposition decreases groove depth if left for too long. This could prevent cell alignment from occurring, particularly when using shallow grooves.

9. Masters should be silanized as necessary, typically every 3–5 times they are used. Based on the inexpensive nature and shallow grooves of the diffraction grating masters, it is probably best to make new ones rather than re-silanizing the used ones.

10. Because of surface tension, the edges of the PDMS will be raised up, and will need to be trimmed later to create a flat surface, so do not worry about keeping them intact.

11. To prevent the PDMS negative from moving, a thin layer of PDMS can be added to the support material before placing the negative on it. This will adhere the negative to the support material.

12. When removing the thin PDMS layer from below a silicon master, be careful not to cut too deeply and damage the mould.

13. The easiest way of ensuring the PDMS sheet fits under the 96-well plate is to mark the cut sites with a marker after placing the sheet feature side up on the bottom side of an upside-down plate. Approximately match the features to desired plate configuration before marking. The sheet should then be trimmed using a scalpel with the edge of a right-side up plate as guide rather than a ruler, to prevent pressing on the features and damaging them.

14. The viscosity of PDMS makes it difficult to draw into the syringe without adding bubbles. Draw the PDMS into the syringe slowly to limit the amount of bubbles. Also, draw in much more than the necessary 0.3 mL. This will allow you to get a better volume measure even when there are large bubbles near the plunger.

15. The plate should not be clamped to the working surface, but rather, to itself so it can be moved to the oven.

16. Although the plasma cleaning step is optional, it will enhance cell adhesion to the substrate.

17. The OrientationJ ImageJ plug-in (16) may be used for auto-mated analysis of cytoskeletal alignment.

18. Noisy images may benefit from the use of a Gaussian blur filter (*Process* menu, then *Filters*, then *Gaussian blur*). By blurring the image, noise becomes less dominant, and the cell membrane is highlighted.

References

1. Wu J, Du Y, Watkins SC, Funderburgh JL, Wagner WR (2012) The engineering of organized human corneal tissue through the spatial guidance of corneal stromal stem cells. Biomaterials 33:1343

2. LeGrice IJ, Smaill BH, Chai LZ, Edgar SG, Gavin JB, Hunter PJ (1995) Laminar structure of the heart: ventricular myocyte arrangement and connective tissue architecture in the dog. Am J Physiol 269:H571–H582

3. Lieber RL, Fridén J (2000) Functional and clinical significance of skeletal muscle architecture. Muscle Nerve 23:1647–1666

4. Carroll TJ, Yu J (2012) The kidney and planar cell polarity. In: Yang Y (ed) Planar cell polarity during development, 1st edn. Academic Press, Oxford

5. Langille BL, Adamson SL (1981) Relationship between blood flow direction and endothelial cell orientation at arterial branch sites in rabbits and mice. Circ Res 48:481–488

6. Vladar E, Bayly R, Sangoram A, Scott M, Axelrod J (2012) Microtubules enable the planar cell polarity of airway cilia. Curr Biol 22:2203

7. Hotulainen P, Lappalainen P (2006) Stress fibers are generated by two distinct actin

assembly mechanisms in motile cells. J Cell Biol 173:383–394

8. Oakley C, Brunette DM (1995) Response of single, pairs, and clusters of epithelial cells to substratum topography. Biochem Cell Biol 73:473–489

9. Oakley C, Brunette DM (1993) The sequence of alignment of microtubules, focal contacts and actin filaments in fibroblasts spreading on smooth and grooved titanium substrata. J Cell Sci 106:343–354

10. Wang C, Baker BM, Chen CS, Schwartz MA (2013) Endothelial cell sensing of flow direction. Arterioscler Thromb Vasc Biol 33:2130–2136

11. Wang JH, Goldschmidt-Clermont P, Wille J, Yin FC (2001) Specificity of endothelial cell reorientation in response to cyclic mechanical stretching. J Biomech 34:1563

12. Rajnicek AM, Foubister LE, McCaig CD (2008) Alignment of corneal and lens epithelial cells by co-operative effects of substratum topography and \DC\ electric fields. Biomaterials 29:2082

13. Teixeira AI, Abrams GA, Bertics PJ, Murphy CJ, Nealey PF (2003) Epithelial contact guidance on well-defined micro- and nanostructured substrates. J Cell Sci 116:1881–1892

14. Londono C, Loureiro MJ, Slater B, Lücker PB, Soleas J, Sathananthan S, Aitchison JS, Kabla AJ, McGuigan AP (2014) Nonautonomous contact guidance signaling during collective cell migration. Proc Natl Acad Sci U S A 111:1807–1812

15. Schneider CA, Rasband WS, Eliceiri KW (2012) NIH Image to ImageJ: 25 years of image analysis. Nat Methods 9:671–675

16. Rezakhaniha R, Agianniotis A, Schrauwen JTC, Griffa A, Sage D, Bouten CVC, Vosse FN, Unser M, Stergiopulos N (2012) Experimental investigation of collagen waviness and orientation in the arterial adventitia using confocal laser scanning microscopy. Biomech Model Mechanobiol 11:461–473

Methods in Molecular Biology (2014) 1202: 57–78
DOI 10.1007/7651_2013_33
© Springer Science+Business Media New York 2013
Published online: 7 September 2013

Bioreactor Cultivation of Anatomically Shaped Human Bone Grafts

Joshua P. Temple, Keith Yeager, Sarindr Bhumiratana, Gordana Vunjak-Novakovic, and Warren L. Grayson

Abstract

In this chapter, we describe a method for engineering bone grafts in vitro with the specific geometry of the temporomandibular joint (TMJ) condyle. The anatomical geometry of the bone grafts was segmented from computed tomography (CT) scans, converted to G-code, and used to machine decellularized trabecular bone scaffolds into the identical shape of the condyle. These scaffolds were seeded with human bone marrow-derived mesenchymal stem cells (MSCs) using spinner flasks and cultivated for up to 5 weeks in vitro using a custom-designed perfusion bioreactor system. The flow patterns through the complex geometry were modeled using the FloWorks module of SolidWorks to optimize bioreactor design. The perfused scaffolds exhibited significantly higher cellular content, better matrix production, and increased bone mineral deposition relative to non-perfused (static) controls after 5 weeks of in vitro cultivation. This technology is broadly applicable for creating patient-specific bone grafts of varying shapes and sizes.

Keywords: Scaffold, Bone, Anatomical, Craniofacial, MSCs

1 Introduction

Bone tissue engineering has the potential to generate defect-specific functional bone grafts for skeletal reconstruction. A successful bone graft would ideally match the mechanical and physiological properties of the implant location and should provide a platform for healing. Autograft is considered to be the gold standard of bone grafting due to its biocompatibility (prevention of immunogenic responses), osteogenicity (containment of osteoblasts and/or osteoprogenitors), osteoconductivity (recruitment of bone-forming cells), osteoinductivity (induction of osteoblastic differentiation), and native mechanical properties. However, several drawbacks related to its use—including donor-site morbidity

---

Joshua P. Temple and Keith Yeager have contributed equally.

and lack of defect-specific size and shape—highlight the need to develop more effective and practical tissue engineering techniques.

Early studies showed that bone tissue may be formed in vitro by growing osteoblastic cells on 3D scaffolds with culture medium containing osteoinductive factors. Subsequent studies demonstrated the potential to induce stem cells from embryonic and adult tissues to differentiate into osteoblasts and form bone tissue in response to biological and mechanical stimuli (1–3). The cells within the engineered bone grafts express osteogenic genes and mineralized extracellular matrix. The increased mineral content and bone-like architectures can increase the mechanical properties of the scaffold and result in tissue constructs with compressive properties approaching those of native bone tissue (4–6). The development of bioreactor systems capable of providing effective nutrient transfer to cells embedded within a scaffold has enabled the formation of large (centimeter-scaled) tissue engineered bone grafts in vitro with homogenously distributed tissue (7–9). With such advancements, our group has demonstrated the ability to engineer anatomically shaped bone grafts from human adult stem cells with clinically relevant sizes by employing 3-dimensional (3D) scaffold fabrication techniques and an advanced cultivation bioreactor system. Engineering bone tissue in these specially designed cultivation systems offers the potential to provide an alternative source of a functional bone graft in lieu of autografts.

A method to successfully engineer anatomically shaped human bone grafts is described in extensive detail in this chapter, including adult stem cell cultivation, scaffold preparation and fabrication, cell seeding, bioreactor assembly, and bone graft cultivation (see Fig. 1). This process covers the technique for fabricating an anatomically shaped scaffold from a trabecular bone block and the design of a perfusion bioreactor system which delivers sufficient nutrients to the cells throughout the scaffold. The construct anatomy and reconstruction processes were adapted from computer-aided surgical planning which is a common process in surgical procedure for complex skeletal anatomy and fabrication of alloplastic grafts. In brief, the anatomically shaped graft was designed by segmenting out the region of interest from the 3D-reconstructed CT images of the patient's skull to replicate the specific geometry. The scaffold was prepared from decellularized, bovine trabecular bone and was machined using a 4-axis CNC milling machine. The perfusion bioreactor culture chamber was fabricated to match the exact shape of the bone graft to ensure a tight seal around the scaffold and ensure medium perfusion through the interstitial spaces within the construct rather than around the periphery. The perfusion pattern through the complex geometry was modeled using the FloWorks flow simulation module of SolidWorks and was used to optimize the bioreactor design. The assembly of the culture system and troubleshooting for commonly occurred malfunctions are explicitly described and the anticipated results discussed.

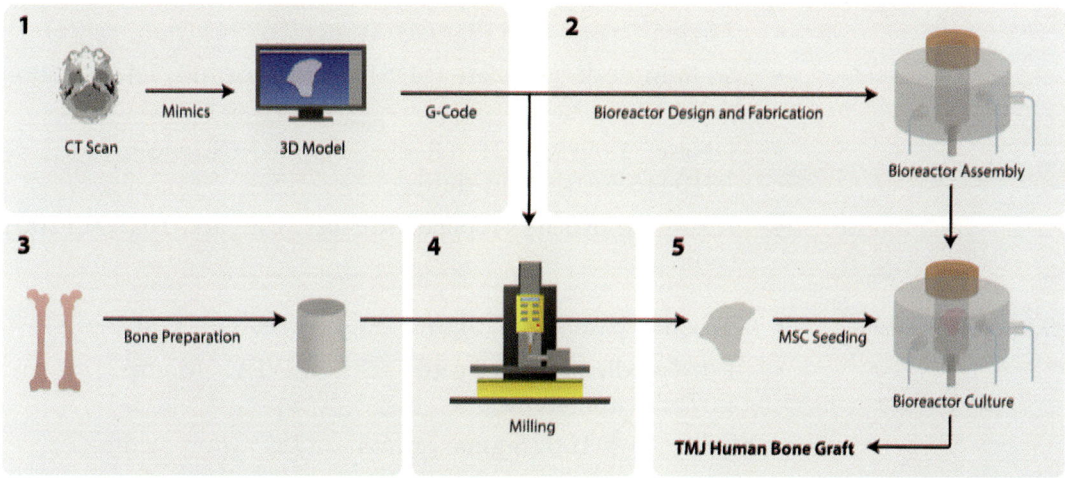

Fig. 1 Schematic of process for generating anatomically shaped human bone grafts. (*Step 1*) 3D reconstructive process used to generate a CAD-compatible 3D model from raw CT scan data. (*Step 2*) CAD model used to design and build customized bioreactors. (*Step 3*) The distal femoral condyles from calf knees are sectioned and decellularized into cylindrical blanks appropriate for milling. (*Step 4*) CNC milling is used to generate scaffolds in the desired anatomical shape. (*Step 5*) The scaffold is seeded with hMSCs and cultured in the bioreactor for 5 weeks

2 Materials

2.1 Reagents

2.1.1 Cell Culture

- High-glucose Dulbecco's modified eagle medium (GIBCO, cat. no. 11995) (sterile).
- Fetal bovine serum, non-heat-inactivated (Atlanta Biological, cat. no. S11550) (sterile).
- Penicillin–streptomycin solution, 100× (Cellgro, cat. no. 30-002-CI) (sterile).
- Recombinant human FGF-basic/FGF-2 (Pepro Tech, cat. no. 100-18B) (sterile).
- β-Glycerophosphate disodium salt hydrate (Sigma, cat. no. G9891) (non-sterile).
- Dexamethasone (Sigma, cat. no. D1756) (non-sterile).
- L-Ascorbic acid-2-phosphate sesquimagnesium salt hydrate (Sigma, cat. no. A8960) (non-sterile).

2.1.2 Decellularization

- 10× Phosphate buffered saline (PBS), pH 7.4 (Quality Biological, cat. no. 119069131) (sterile).
- Ethylenediaminetetraacetic acid (EDTA) (Sigma, cat. no. E6758) (non-sterile).

- Trizma® base (Tris) (Sigma, cat. no. T6791) (non-sterile).
- Sodium dodecyl sulfate (SDS) (Sigma, cat. no. L4390) (non-sterile).
- DNase I grade II (Roche Applied Sciences, cat. no. 10104159001) (non-sterile).
- RNase A (Roche Applied Sciences, cat. no. 10109142001) (non-sterile).

2.1.3 DNA Extraction and Quantitation

- Trizma® base (Tris) (Sigma, cat. no. T6791) (non-sterile).
- Ethylenediaminetetraacetic acid (EDTA) (Sigma, cat. no. E6758) (non-sterile).
- Triton™-X 100 (Sigma, cat. no. T8787) (non-sterile).
- Proteinase K (Sigma, cat. no. P2308) (non-sterile).
- Quant-iT™ PicoGreen® dsDNA Assay Kit (Invitrogen, cat. no. P7589).

2.2 Reagent Setup

2.2.1 Stock Solutions

- Human Basic Fibroblast Growth Factor (bFGF): Prepare at least 10 mL buffer for reconstitution consisting of 5 mM Tris and 0.1 % (wt/vol) BSA, sterile filter and store at 4 °C until use. Add 50 µg of lyophilized FGF-2 and centrifuge vial at $12,000 \times g$ for 1 min. Reconstitute with 1 mL of buffer. Transfer contents to a 15 mL conical tube and add 9 mL of buffer. Split into 50 µL aliquots and store at −20 °C. Thawed aliquots can be stored at 4 °C for up to 1 week.
- β-Glycerophosphate: 200 mM β-glycerophosphate dissolved in high-glucose DMEM, sterile filtered, aliquoted, and stored at −20 °C.
- Dexamethasone: 1 mM dexamethasone dissolved in 100 % ethanol, sterile filtered, split into 500 µL aliquots, and stored at −20 °C.
- Ascorbic Acid-2-Phosphate: 5 mM ascorbic acid dissolved in high-glucose DMEM, sterile filtered, split into 1 mL aliquots, and stored at −20 °C.

2.2.2 Cell Culture Media

- Expansion medium: 10 % (vol/vol) FBS, 1 % (vol/vol) pen–strep, and 0.1 ng/mL bFGF. Expansion media should be stored at 4 °C. Once supplemented with bFGF, medium can be used for up to 1 week.
- Osteogenic differentiation media: 10 % (vol/vol) FBS, 1 % (vol/vol) pen–strep, 10 mM sodium-β-glycerophosphate, 1 nM dexamethasone, 50 µM ascorbic acid-2-phosphate with the remaining volume made up with high-glucose DMEM. Osteogenic media should be made fresh weekly and stored at 4 °C.

2.2.3 Bone Decellularization Solutions

- Detergent wash 1: 0.1 % EDTA (wt/vol) in 450 mL deionized water and 50 mL $10\times$ PBS. Stored at room temperature.

- Detergent wash 2: 0.1 % EDTA (wt/vol) and 10 mM Tris in 500 mL deionized water. Stored at room temperature.

- Detergent wash 3: 10 mM Tris and 0.5 % SDS (wt/vol) in 500 mL deionized water. Stored at room temperature.

- Enzymatic wash: 50 units/mL DNase, 1 unit/mL RNase, and 10 mM Tris in 450 mL deionized water. Stored at 4 °C for up to 1 month.

2.2.4 DNA Extraction and Quantitation

- DNA extraction solution (TEX + proteinase K): 10 mM Tris, 1 mM EDTA, 0.1 % Triton-X, and 0.1 mg/mL proteinase K stored at −20 °C.

- DNA quantitation: Following protocol for Quant-iT™ Pico-Green® dsDNA Assay Kit with a cell solution with known cell number to determine the amount of DNA per cell.

2.3 Equipment

- Transparent vacuum desiccator (Thermo, cat. no. 53110250).
- 16-G Luer-Lok™ needles (Fisher, cat. no. 1482618A) (sterile).
- 4 oz. screw-top polypropylene histology container (Fisher, cat. no. 22026310) (sterile).
- Incubator.
- Biological safety cabinet.

2.4 Bone Segmentation

- Bandsaw appropriate for segmenting bone and cutting bioreactor components. (Optional) Water-cooled bone bandsaw (Mar-Med, cat. no. 80) reduces dust and splatter.

- Bandsaw blades: it can be useful to have multiple as they tend to break. (Optional) Diamond bandsaw blade (Mar-Med, cat. no. 75) will not cut hands and easily slices bone.

2.5 Bioreactor

- Screw-top polypropylene histology containers (Fisher, cat. no. 22026311) (non-sterile).
- Three-prong swivel ring stand clamps (Fisher, cat. no. 02300209).
- Disposable 20 mL Luer-Lok™ syringes (Fisher, cat. no. 148232B) (sterile).
- 100 mL Glass media bottles with cap (Fisher, cat. no. FB800100) (non-sterile).
- 20″ Ring stand (Fisher, cat. no. 14675BQ).
- 26-G Luer-Lok™ needles (Fisher, cat. no. 1482610) (sterile).
- 23-G, 1″ length Luer-Lok™ needles (Fisher, cat. no. 14826A) (sterile).

2.6   Milling
- Starrett 0.2″ tip edge-finder (McMaster-Carr, cat. no. 20535a653).
- 3/16″ dia. solid carbide ball endmill (MSC Industrial Supply, cat. no. 07766645).
- 3/16″ dia. screw machine length twist drill bit (McMaster-Carr, cat. no. 2908A39).
- 1 ft long, 25 mm dia. white Delrin® acetal plastic rod (McMaster-Carr, cat. no. 8572 K61).
- Anti-fog, anti-scratch safety glasses (MSC Industrial Supply, cat. no. 89972509).
- Shop-Vac 6-gal 3-HP wet/dry vacuum (Aubuchon Hardware, cat. no. 118208).
- (Optional) 8″ Black oxide hand file (McMaster-Carr, cat. no. 42405A45).
- (Optional) High-pressure precision compressed air can (McMaster-Carr, cat. no. 8431 K22).
- (Optional) USB jog dial (LittleMachineShop .com, cat. no. 3414).

2.7   Software
- Mimics Innovation Suite (Materialise).
- SolidWorks (Dassault Systèmes).
- Mastercam (Mastercam).
- Mach3 (ArtSoft USA).

3   Methods

3.1   3D Reconstruction from CT

1. Open Materialise Mimics and select File > Open. Navigate to desired DICOM directory file and import it. Follow the software prompts to setup the file and orient the slices.

2. Select *Segmentation* > *Thresholding* and select the *Bone* preset from the dropdown menu. Click *OK*.

3. Scan through the slices to make sure the mask is correctly selecting bone. Adjust the threshold cutoffs from the previous menu if necessary.

4. The *Multiple Slice Edit* and *Edit Mask* tools (accessed from the *Segmentation* menu) can add or remove areas of the mask and separate areas into new masks if necessary.

5. Select *Segmentation* > *Calculate 3D*, choose *Optimal* under *Quality* and click OK to generate a 3D model of the selected mask. Depending on the computer used to run Mimics, you may need to adjust the *Quality* settings.

6. To isolate a specific area of the scan as a model (in this case, the temporomandibular joint), select *Segmentation > Edit Mask in 3D*. Mimics will display a white bounding box on each mask viewpoint. Adjust this bounding box to crop the mask rendering. In the 3D window, draw around the area to be isolated. It will change color.

7. Select the *Separate* button to create a new mask from the selected area. A new model can be generated from this cropped mask in the same way as in step 5.

8. (Optional) Right clicking on the model name in the *3D Objects* tab and selecting properties

9. The *Smoothing*, *Triangle Reduction*, and *Wrap* features (accessed from the Tools menu) are useful for cleaning up the geometry of the model. Avoid overusing these features as they modify the geometry, decreasing the model's correlation to the original scan data.

10. Select *Export > Binary STL*. Under the 3D tab, select the model to be exported and click the *Add* button. Select Finish to generate a .stl file of the model.

3.2 CNC Toolpath Generation (Fig. 2)

1. Open SolidWorks and import the STL file from previous step.

2. Insert a reference plane on the base of the TMJ, coplanar with the axial segmentation plane of the anatomy (Fig. 2a).

3. Insert a sketch and draw a circle of diameter sufficient to enclose the entire silhouette of the TMJ, with its center coincident with the centroid of the silhouette (Fig. 2b).

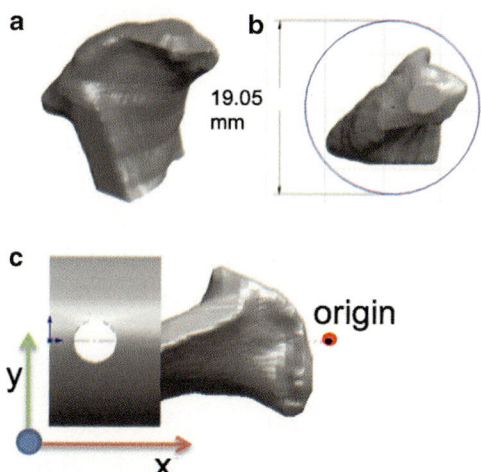

Fig. 2 TMJ preparation for CAM. (**a**) Imported TMJ. (**b**) *Steps 13–14*, adding a cylindrical support base. (**b**) *Steps 15–16*, setting the origin

4. Exit the sketch and create an extruded base feature extending at least 1 cm in the inferior to the base of the condyle (in anatomic orientation) (Fig. 2c).

5. Orient the modified TMJ such that the axis of the cylindrical extruded base is coaxial with the x-axis of the world coordinate system, with the TMJ anatomy along the positive direction (Fig. 2c).

6. Insert a sketch on either the top or front reference plane. Sketch a single point along the x-axis approximately 1 mm to the right of the most positive extent of the geometry.

7. Exit the sketch and move the modified TMJ such that the sketched point is coincident with the origin of the world coordinate system.

8. Measure and record the distance from the origin to the start of the cylindrical feature, as well as the diameter of the cylinder. These parameters are required for later fabrication steps.

9. Export the modified TMJ in IGES format and open in Mastercam.

10. Select the milling machine type from the menu that is associated with the CNC controller installed on your machine.

11. In order to machine geometry accurate to the 3D model, maximum rigidity during the milling process is desired. To achieve this, the toolpaths are subdivided into sections progressing from the material region furthest from the fixture to the regions closest to the fixture. For the human TMJ, using two sections has proven sufficient. The subsequent steps define these regions and the toolpaths associated with them:

12. Orient the view to the right side, looking at the YZ plane. Sketch a circle centered on the x-axis at a negative x value equal to half the axial anatomic length and with a diameter approximately double the bulk diameter of the TMJ.

13. Extrude the circle in the positive x direction at least 5 mm further than the most positive point on the 3D model to create a cylinder defining a tool containment boundary.

14. Create a rotary 4-axis toolpath using a 3/16″ ball endmill at a feed rate of 9 IPM (inches per minute) and 3000 RPM. Use the TMJ surface to generate the toolpath, and use the cylinder from the previous step to establish the tool containment boundary. For this initial toolpath, leave approximately 0.020″ on the drive surface. The 4-axis toolpath should be setup as an axial cut with 10° increments for the full 360° rotation.

15. Duplicate the 4-axis toolpath created in the previous step, and mill the remaining material (remove the 0.020″ offset and set

to zero). Modify the toolpath to have an axial cut with 2° increments.

16. Orient the view to the right side, looking at the YZ plane. Sketch a circle centered on the *x*-axis at a negative *x* value equal to half the axial anatomic length plus 1/4″ (a value at least the size of the cutting tool) and with a diameter approximately double the bulk diameter of the TMJ.

17. Extrude the circle in the negative x direction at least 5 mm further than the most negative point on the anatomic surface to create a cylinder defining a second tool containment boundary.

18. Duplicate the existing toolpaths (the original 4-axis toolpath and its modified duplicate) and modify each of the two new toolpaths to use a containment boundary defined by the cylinder created in the previous step.

19. Generate G-code based on the entire set of toolpaths using a post-processor appropriate for your milling machine. Save the text file generated in an appropriate format (e.g., .txt or .nc) and transfer to your milling machine's controller.

3.3 Bioreactor Design and Optimization

3.3.1 Bioreactor Design

1. In SolidWorks, open the STL file of the TMJ anatomy.

2. As a separate part, model a syringe needle with a standard gauge. In this case, 26G needles were chosen.

3. Create an assembly of the scaffold, and add multiple needles to serve as perfusion ports. We chose three ports arrayed uniformly in the radial orientation, and anatomically targeted in the axial direction.

4. The needle ports are designed so that media exits the scaffold through the needles, and enters the scaffold axially from the lowermost section of the geometry.

5. The needles are targeted in such a way as to allow the entire scaffold to be perfused, and are placed at extremities of the scaffold (medial, lateral, and superior extents).

6. As a separate part, model a syringe casing that is of sufficient size to enclose the entire scaffold, plus an additional 3 cm in length. Do not model the plunger—only the casing plus syringe port is needed. Add a 3/16″ hole in the radial direction that cuts through the entire case (both sides). This hole should be approximately 1 cm from the top of the syringe casing (the end opposite of the Luer-Lok port).

3.3.2 Mold Assembly

1. As a separate part, model a tube with the same inner and outer diameters of the syringe casing, and with a length at least three times the length of the TMJ. Add two 3/16″ holes along the

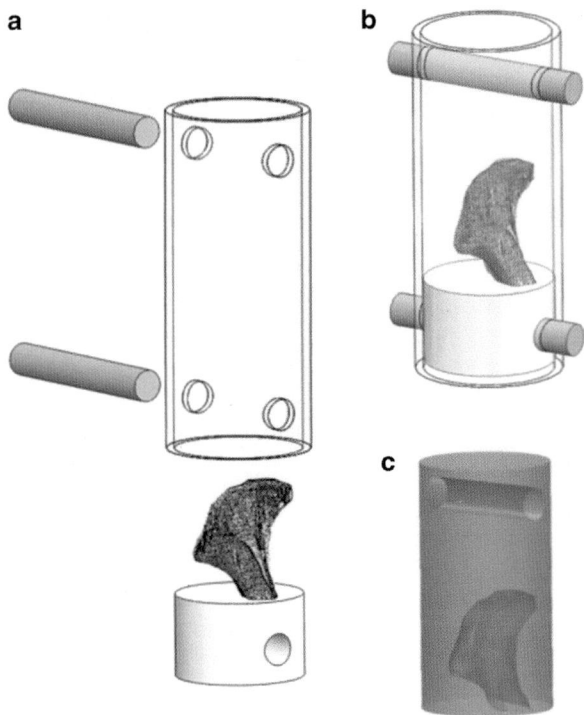

Fig. 3 PDMS mold design. (**a**) *Step 1*, outer wall of mold. *Step 2*, TMJ component with alignment hole. *Step 3*, alignment pins. (**b**) *Step 4*, assembled mold cavity. (**c**) *Step 6,* CAD model of PDMS mold created via boolean subtraction

length of the tube that cut radially through the entire tube, and are spaced approximately twice the length of the TMJ, and centered along the length of the tube (Fig. 3a).

2. As a separate part, duplicate the TMJ scaffold with the 1 cm long cylindrical base (from step 14) with a diameter equal to the inner diameter of the syringe tube. Add a 3/16″ hole in the radial direction in the center of this cylindrical feature

3. As a separate part, model a cylinder of 2″ length and 3/16″ diameter.

4. Create an assembly from two of these cylinders, the scaffold with cylindrical feature, and the tube created in previous steps. Mate one of the 3/16″ holes in the tube with the same size hole in the modified scaffold and mate this combination with one of the cylinders. Create a cylindrical mate to align the modified scaffold within the center of the tube, such that the TMJ geometry is contained within the tube. Align the remaining 3/16″ diameter cylinder with the remaining hole in the tube (Fig. 3b).

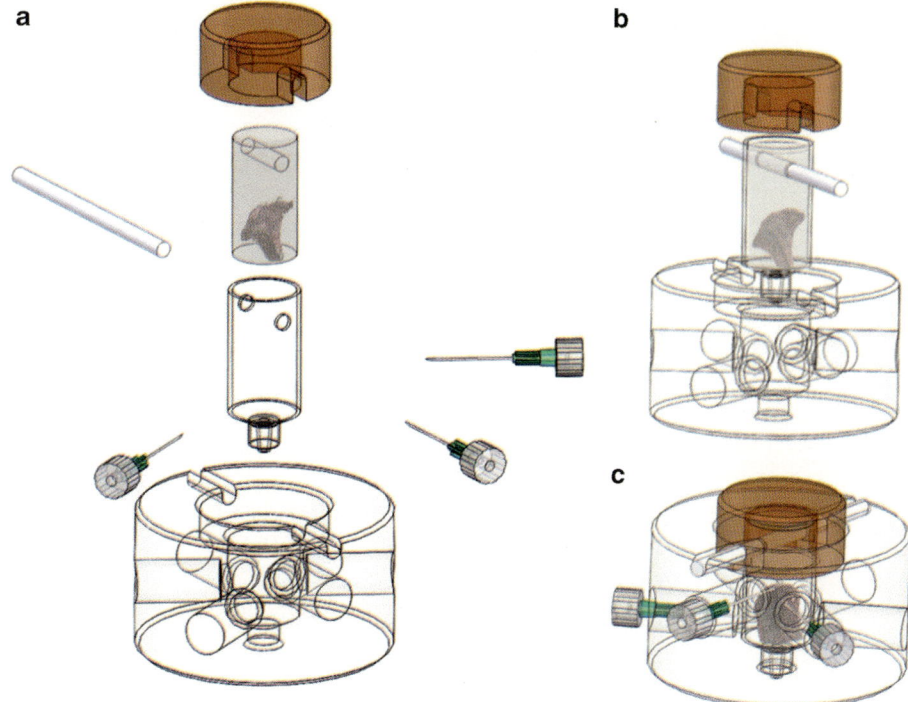

Fig. 4 Bioreactor assembly in solidworks. (**a**) Assembly components previously modeled and newly modeled in *step 3* (needle guide) and *step 4* (syringe casing cap). (**b**) *Step 2*, aligned TMJ, PDMS casting, and alignment pin. (**c**) Fully assembled bioreactor

5. The assembly created in the previous steps represents a mold in which to cast a PDMS component, which will be used to surround the TMJ bone scaffold. For purposes of modeling flow rates during perfusion, this PDMS component will need to be modeled in SolidWorks.

6. Insert a new part into the mold assembly and create a cylinder with diameter equal to the inside diameter of the tube and of a length to match the tube. Perform a Boolean subtract operation on this assembly to create the PDMS component. The result should show a TMJ shaped cavity within a cylinder with a second cylindrical hole above the TMJ running in the radial direction.

3.3.3 Bioreactor Assembly

1. Open the bioreactor assembly and insert the PDMS component, the syringe casing, and one 3/16″ cylindrical component, which will be used as an alignment rod (Fig. 4a).

2. Align and mate the scaffold within the PDMS cavity. Next, align the PDMS cavity to be internal and concentric with the syringe casing. Align the 3/16″ hole in the PDMS cavity with the 3/16″ hole in the syringe casing and mate the 3/16″ diameter alignment rod to pass through these holes (Fig. 4b).

3. Create a new part and model the housing that supports and surrounds the syringe casing assembly, leaving the top open to the alignment rod. Cut a channel in the top of the housing for the alignment rod. Cut holes in a radial orientation aligned with each perfusion needle (Fig. 4a–c).

4. Create a new part and model a cap that surrounds the top of the syringe casing and extends slightly below the alignment rod. Cut a channel in the cap for the alignment rod. For the inner diameter of the cap that interfaces with the syringe casing, decrease the diameter by approximately 0.5 mm to achieve a press fit during the assembly process (Fig. 4a–c).

3.3.4 Fluid Modeling

1. Create a new fluid simulation on the bioreactor assembly.

2. Edit the PDMS part and remove the material where the internal diameter of the needle intersects the PDMS. There should be a channel from the outside of the PDMS to the internal TMJ cavity where each needle is placed.

3. Modify each needle and the bottom of the syringe casing (the port) and model a plug on each opening to close the tube.

4. The faces of these plugs are used to specify boundary conditions for the fluid modeling. In the fluid simulation setup create an inlet flow rate of 1 mL/min at the syringe port. For each needle port opening, model this boundary interface as open to ambient pressure

5. In the fluid model, define the scaffold as a porous medium. (A value of 0.7 represents 70 % porosity.) The fluid viscosity should be adjusted to represent the media perfused through the scaffold. In our modeling, we used water as the fluid.

6. Run the fluid model and create a velocity map on several cross sections of the scaffold (Fig. 5a, b). The velocity will always increase near the needle port exits. There will also be variations in flow direction and velocity due to the complex geometry of the graft. However, flow should be optimized to be fairly uniform throughout the bulk of the scaffold. To adjust the perfusion through the scaffold, needles can be repositioned, more can be added, or their sizes can be changed to achieve the desired flow throughout the scaffold.

7. Once the flow model has been refined, save the final geometry and fabricate the needle guide based on the model's results (Fig. 5c).

3.4 Bioreactor Fabrication

3.4.1 Component Construction

1. Alignment rod: cut a 3/16″ stainless steel, alloy 316 rod to length using a bandsaw and de-burr the edges.

2. Syringe casing: cut a 30 mL syringe to length using a bandsaw and drill a 3/16″ hole at specified location using a vertical milling machine. You may also wish to pre-drill holes for

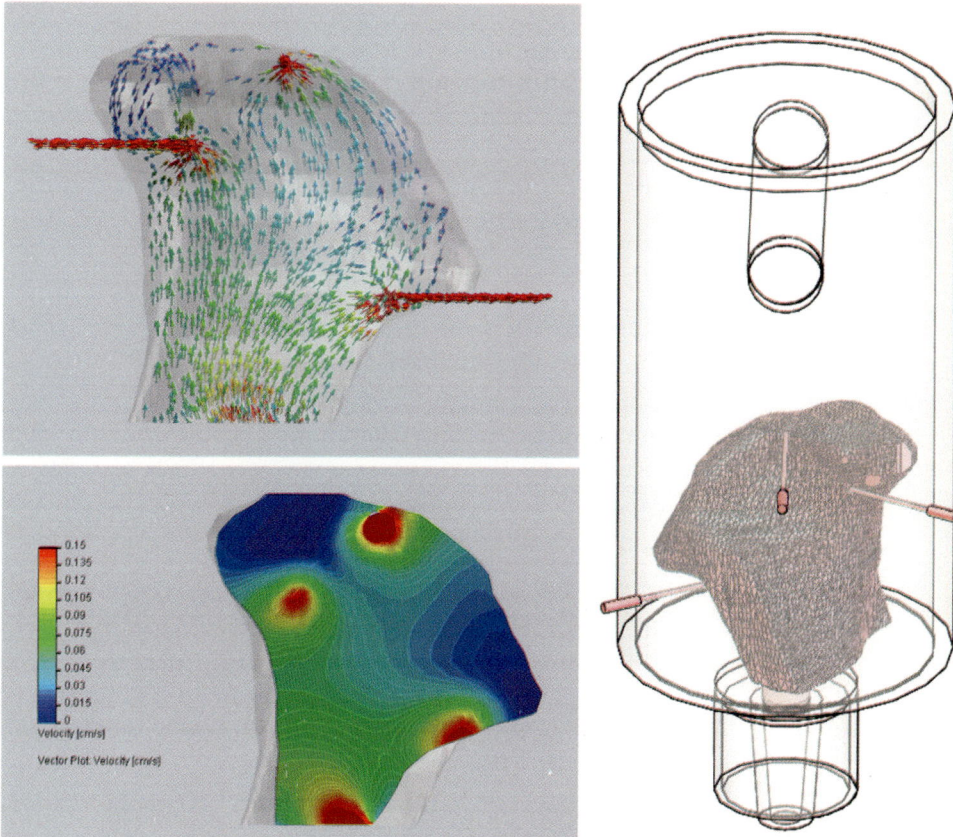

Fig. 5 Modeling flow to optimize bioreactor design. (*Top left*) Color-coded velocity vectors of flow through scaffold. (*Bottom left*) Cross-section through the center of the scaffold gives spatial distribution of medium flow using color-coded scalar values. Computer-aided design of outlet port placement based on optimization of flow patterns throughout the scaffold

needles to enter the casing more easily. For this, you will need a hexagonal collet block, and a collet of appropriate size to hold the syringe casing. This allows the user to drill small pilot holes at 60° angular increments on the length. For other angles, a fourth axis is needed, which can be manually operated or automated.

3. PDMS mold: see next section.

4. Cap: this component is made from polyetherimide and machined to specifications using a lathe. The component is then secured on a milling machine to cut the channel for the alignment rod.

5. Housing: the concentric cylindrical cuts of this component are made first on a lathe in acrylic or polycarbonate plastic. The syringe-case-port end of the housing should have an additional 1″ diameter extrusion machined about 0.5″ length to serve as a

fixing feature for holding in a collet block or fourth axis collet for machining on a vertical milling machine. On the milling machine, drill the holes needed for the perfusion needles, as well as the channel for the alignment rod.

3.4.2 PDMS Mold Pouring

1. Remove the plunger from a 20 mL syringe. Cut off the Luer-Lok® end 2″ from the tip. The Luer-Lok® end will be used in the bioreactor, while the remainder will serve as the mold casing.

2. Using a 3/16″ twist drill bit, drill two holes, 1.5″ apart from each other, through both sides of the syringe.

3. Fit the drilled syringe casing over the TMJ mold machined from acetal plastic and insert a metal rod through each hole.

4. Weigh out 12.36 g of the PDMS base mixture, tare the balance, and add 1.236 g of the curing agent. Mix components well (Note 1).

5. Transfer the mixture to a vacuum desiccator and degas until there are no bubbles visible in the mixture (30–60 min).

6. Remove mixture from the vacuum and pour the PDMS into the syringe casing smoothly and quickly to avoid incorporating more air. Return to the vacuum desiccator until bubbles are gone (up to 1 h).

7. Meanwhile, preheat the oven to 80 °C. Transfer mold to oven and bake for 30–45 min.

8. To remove the PDMS from its casing, the syringe must be cut open. Using wire cutters, carefully cut into the plastic and gradually peel and cut it away from the PDMS. Next, make a small incision along the length of the TMJ and pull the PDMS off of the acetal mold.

3.5 Bone Blank Preparation

1. Using a sharp surgical blade, separate the femur and tibia of the knee of 2–4 month old calves by severing the ligaments. Remove the meniscus and clean as much connective tissue and fat from around the joint as possible. The distal femoral condyle generally yields the more uniform trabecular bone than the tibia.

2. Using a bandsaw, cut through the femur about 10 cm above the condyle and discard the diaphysis.

3. Shave off a small slice of bone (about 1 cm thickness) along the length of the section to reveal the growth plate. Slice the remainder into a block with dimensions roughly approximating $8 \times 4 \times 4$ cm with an adequately large section of trabecular bone not split by the growth plate. The same process can be repeated to varying degrees of success (due to smaller size) with the tibial condyle.

4. Depending on the lathe being used, you may need to cut away the thickness on the end of the block to facilitate firm grasping by the lathe chuck jaws.

5. Clean the block with a high velocity water jet. It will not be possible to remove all marrow from the inside of the block, but washing at this stage reduces splatter while turning.

6. Secure the block in the lathe chuck, tightening enough to keep the block snug, but taking care not to crush or warp the block. Center the block axially within the chuck so it does not wobble heavily while spinning (Note 2).

7. Turn on the lathe and begin cutting at the widest portion of the bone, sweeping across the bone horizontally. After each pass, increase the cutting depth by 1 mm. Stop cutting once the bone cylinder reaches a diameter of 25 mm.

8. Clean the cylinder with a high velocity water jet, wrap in aluminum foil, and freeze at −20 °C to store.

9. Cleaned bone blank cylinders may be stored in aluminum foil at −20 °C for up to 3 months.

3.6 TMJ Scaffold Decellularization

1. Place scaffolds in detergent solution 1 and place on rocker for 1 h at room temperature (18–25 °C).

2. Repeat previous step, but in detergent solution 2 for at least 12 h at 4 °C.

3. Remove scaffolds from detergent and transfer to PBS for 1 h at room temperature (18–25 °C).

4. Transfer scaffolds to detergent solution 3 and place on rocker for at least 24 h at room temperature (18–25 °C).

5. Place scaffolds in fresh PBS and agitate for 1 min. Drain PBS and replace with fresh PBS. Continue washing until bubbles do not form in the PBS when agitated (this is a sign that all detergent has been removed).

6. Transfer scaffolds to enzymatic wash solution and place on rocker for 5 h at room temperature (18–25 °C).

7. Rinse scaffolds extensively first with PBS and then with deionized water.

8. Prior to use in culture, place in 100 % ethanol overnight on the rocker at room temperature (18–25 °C) (Note 3).

3.7 Scaffold Machining

1. Set Mach3 to millimeters using the Settings tab.

2. Fourth-axis machining is performed using a rotary table mounted on the mill, providing the A axis in addition to the XYZ axes. Place the rotary table on the mill table with the chuck facing along the X-axis and fasten it loosely with the corresponding bolts.

3. Load a dial indicator into the collet of the mill. This indicator will be used to align the rotary table parallel with the axis of the mill.

4. Step the indicator up to a corner of the rotary table, using the dial to find the precise edge.

5. Jog the indicator along the edge of the rotary table to the opposite side. If the indicator shows pressure, lightly tap the rotary table away from the indicator with a rubber mallet until the dial returns to zero. If no pressure is shown, step the indicator up to the rotary table's exact edge and repeat this step.

6. Tighten the bolts securing the rotary table, but not completely. Repeat steps 4 and 5 and tighten down the bolts all the way.

7. Secure the bone cylinder in the rotary chuck, seeking to align it horizontally along the X-axis. Avoid heavily crushing the bone while tightening.

8. Now that the workpiece is secured, you will need to define the machining origin $(0, 0, 0)$. This point will be located in the center of the bone cylinder at the end farthest from the rotary table. Load an edge-finder into the collet of the mill. Slightly displace the tip of the edge-finder from its central axis and start the spindle at a low RPM (300–600).

9. Step the edge-finder up to the edge of the cylinder that extends along the X-axis. As the edge-finder contacts the edge of the workpiece, the workpiece will gradually center the tip of the edge-finder. When the tip is spinning precisely in alignment, the machine is half the radius of the tip from the workpiece.

10. In Mach3, select the Y-axis coordinate readout and type in the distance from the central axis of the cylinder. This distance is the radius of the edge-finder tip (in millimeters), added to the radius of the cylinder.

11. To zero the X-axis, repeat step 9, using the face of the cylinder as the edge.

12. Select the X-axis coordinate readout in Mach3 and type in the distance from the face of the cylinder. This distance is the radius of the edge-finder (in millimeters). This number should be negative.

13. To zero the Z-axis, load the 3/16″ ball endmill into the collet of the mill.

14. Step the Z-axis down until there is a slight drag on a sheet of copy paper slid between the workpiece surface and the cutting tool. This point is approximately 0.1 mm above the workpiece.

15. Select the Z-axis coordinate readout in Mach3 and type in the radius of the workpiece added to the thickness of the paper (0.1 mm). All zeros should now be defined (Note 4).

16. Press Cycle Start to begin the program (Note 5).

3.8 Scaffold Cell Seeding and Incubation

1. Scaffolds should be sterilized in 70 % EtOH overnight prior to cell seeding.

2. Remove the lid lining from one seeding container and loosely cap it with the lid. Autoclave the container and a magnetic stirrer bar of appropriate length.

3. To construct the seeding container, make holes in the lid of the plastic with a 16-G needle, three holes per scaffold to be seeded.

4. Expand MSCs to a cell count of 30 million and resuspend in 30 mL of expansion medium (1×10^6 cells/mL).

5. Under sterile conditions, securely attach the scaffold to the container lid needles.

6. While remaining sterile, add the stirrer bar to the seeding container and transfer the cell suspension to the container.

7. Invert the lid and submerge the scaffolds in the cell suspension. Place the container on a magnetic stirrer at 300 RPM.

8. Allow the scaffold to seed for 1 h in the incubator.

9. Remove the scaffold from the seeding container and transfer it to a 50 mL tube containing 10 mL of osteogenic medium. Allow the scaffolds to culture statically for 7 days in the incubator.

10. To determine seeding density, a scaffold is taken from the conical tube 1 day after seeding. DNA is extracted in 5 mL TEX with 0.1 mg/mL proteinase K overnight at 56 °C and the total DNA is quantified with Quant-it™ PicoGreen® assay kit. Cellular density can be determined based on a DNA extraction of a known cell number and total scaffold volume base on 3D image reconstruction.

3.9 Bioreactor Assembly

1. Place bioreactor components (syringe casing, PDMS gasket, alignment rod, cap, and needle guide), tubing (with attached needles and screws), media reservoir, and forceps into autoclave pouches and sterilize using the "liquid cycle," 121 °C (see Fig. 6 for bioreactor components and assembled bioreactor).

2. Assembly requires two people: One who handles the sterile parts of the assembly, and one who prepares the components. The sterile person should put on sterile gloves before proceeding. Throughout the process, the non-sterile person will open autoclave pouches, handle the outsides of components, and position tools for the person doing the sterile assembly.

3. Remove the forceps from their autoclave pouch and place into a 50 mL conical tube filled with 70 % EtOH. This will keep the forceps clean while not in use.

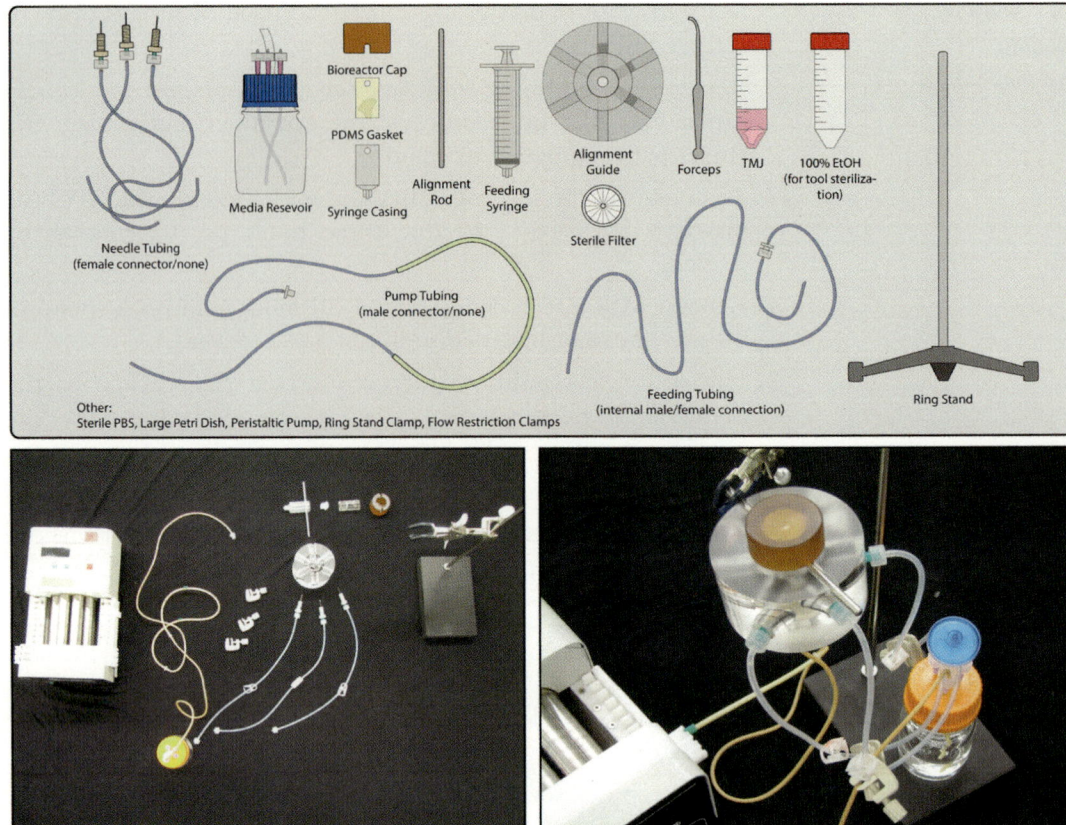

Fig. 6 Bioreactor assembly. (*Top*) Listing of all bioreactor components. This list should be consulted prior to sterilization and assembly to ensure that all components are present. (*Bottom left*) Exploded view of each bioreactor component. Parts can be sterilized partially assembled in this fashion in separate autoclave pouches. (*Bottom right*) Fully assembled TMJ bioreactor

4. Fill the lid of a large sterile petri dish with sterile PBS. If at any time, a component needs to be set down, it can be placed in this dish. In addition, the PBS will aid in inserting the PDMS gasket into the syringe casing.

5. Remove the PDMS mold from its pouch.

6. Remove the forceps from the ethanol and let dry for a few seconds. Since ethanol is toxic to cells, it is important to let it evaporate.

7. Using the forceps, remove the TMJ scaffold from the media and place it inside the PDMS gasket.

8. Remove the syringe casing from its pouch and slide the PDMS gasket into it, taking care to align the holes while inserting. Rolling the PDMS in the sterile PBS may help lubricate the gasket for insertion. This step requires a significant amount of force to fit the gasket into its casing.

9. Insert the alignment rod into the hole in the syringe casing. Leave enough of the rod on one side to properly clamp the bioreactor later.

10. Place the syringe casing into the needle alignment guide, lining up the alignment rod with the grooves on top of the guide.

11. Press fit the bioreactor cap on top of the assembly and set the bioreactor down onto the Petri dish.

12. Remove the media reservoir from its autoclave pouch. Place a sterile filter into the thick tubing to allow for air exchange. Fill the reservoir with 80 mL of osteogenic medium.

13. Remove the needle tubing one-at-a-time from the autoclave pouch and connect the open end of the tubing to the media reservoir.

14. Disconnect the tubing from the needle end and hand the needle to the non-sterile assistant. Thread the screw into the guide, taking care that the needle does not deflect up or down. Thread the screw all the way in, and verify that the needle penetrates the scaffold.

15. Reconnect the tubing. Repeat step 14 with the remaining two needle tubes.

16. Remove the pump tubing and connect the open end to the media reservoir. Place the thick section of the tubing into the peristaltic pump cassette. Be sure to line up the cassette arrow with the proper direction of flow. Attach the cassette to the pump, again verifying that the flow proceeds through the base of the bioreactor and out the needles. Do not connect to the bioreactor yet.

17. Turn on the pump at a low speed and wait until media almost reaches the other end of the pump tubing. At this point, stop the pump. This step ensures that air will not be pumped through the scaffold.

18. Tightly connect the other end of the pump tubing to the bioreactor's bottom port.

19. Connect the open end of the feeding tube to the media reservoir and a 10 mL syringe to the opposite end of the feeding tube.

20. Turn on the pump and watch the needle tubes for flow. If there is no flow through all three needle tubes, see Note 6.

21. Watch the timing of the drops of media into the reservoir, and use the flow restriction clamps to equalize the flow rate through each tube so the drops emerge into the reservoir at the same rate.

3.10 Bioreactor Culture

1. During culture, medium should be changed every 3 days. Remove the feeding syringe from the incubator and transfer it to the biological safety cabinet.

2. Use the syringe to remove and subsequently replace 40 mL of medium in the reservoir.

3. Wrap the bioreactor in fresh Parafilm and return to the incubator.

3.11 Anticipated Results

3.11.1 Cellular Density

The seeding efficiency calculated after 1 day after seeding was approximately 34 %, resulting in a cellular density of approximately 3.4×10^6 cells/mL of tissue. The cultivation of the scaffold in static culture for 1 week allowed the cells to become firmly attached to the scaffold prior to exposing cells to perfusion. Cells proliferated extensively over the first week of culture, as evidenced by an approximately 7.5-fold increase in DNA content. Cultivation in perfusion bioreactor further increased the DNA content by an approximately 75-fold increase relative to the initial seeding value as opposed to an only 37-fold increase in DNA content when cultured in static condition. As a result, the total cellular density in the scaffold under perfusion culture was approximately 250×10^6 cells/mL tissue. Histological analysis showed homogenous cell distribution throughout the construct (Fig. 7a).

3.11.2 Bone Matrix Deposition

Over the cultivation period, MSCs differentiated and deposited new tissue throughout the entire tissue volume. In addition to cellularity, histological sections demonstrated osteoid formation patterns throughout the constructs (Fig. 7b). SEM images showed pore spaces that were densely packed by cell and matrix structure (Fig. 7c). The increase in mineral content was also evident from the 3D reconstructions of μCT images (Fig. 7d, e). After 5 weeks of cultivation, the bone volume in constructs grown in the perfusion bioreactors increased by approximately 11.1 %.

4 Notes

1. Both the base mixture and curing agent have low viscosities. Pour slowly to avoid overshooting.

2. Turning bone will splatter marrow over large distances. It is advisable to cover as much of the body, face, and surrounding area as possible when performing this step. Freezing the bone block immediately prior to lathing may also reduce splatter.

3. Decellularized scaffolds should be lyophilized prior to storage. Clean, lyophyllized scaffolds can be stored at 4 °C indefinitely.

4. It is generally best to "cut air" before running an unfamiliar program to make sure nothing unexpected is taking place in the code. To do this, jog the machine to a position safely above the part, set the Z-axis location to zero, turn off the spindle, and then run the program.

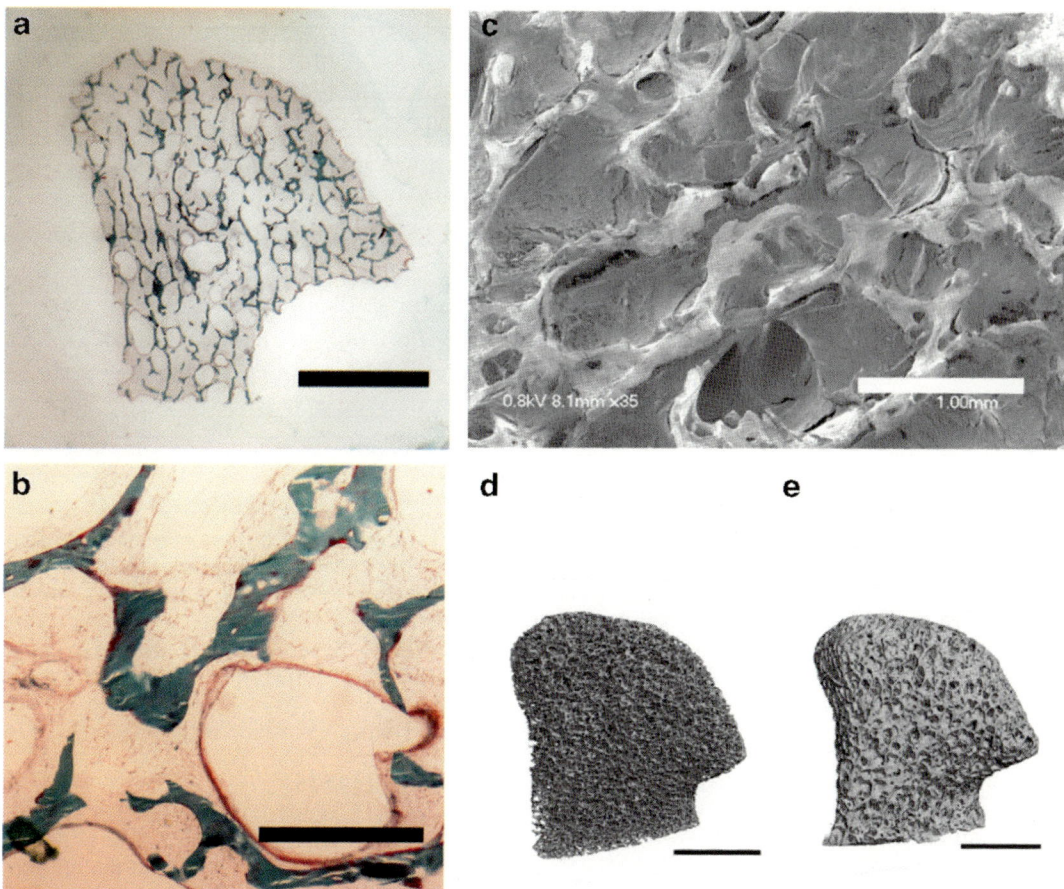

Fig. 7 Representative data of bone grafts cultured for 5 weeks in bioreactor. (**a**) Histological section of plastic embedded bone tissue section stained with Goldner's Trichrome. *Green* stain indicate bone scaffold. *Red* indicates new tissue formation. (**b**) Higher magnification image of histological section demonstrating new tissue formation in the interstitial spaces of the scaffold. *Dark red* stains indicate osteoid formation. (**c**) SEM image of central region of bone graft showing dense tissue formation following bioreactor cultivation (Scale bar = 1 mm). Micro-CT image of unseeded scaffold (**d**) and graft cultured in the bioreactor for 5 weeks (**e**) (scale bars = 5 mm)

5. Avoid sending the machine to (0, 0, 0) as the origin is now defined inside the part. Load the TMJ G-code file in Mach3 and check the tool path preview to ensure that the controller is configured properly.

6. Flow through the needle tubes may be prevented by a number of reasons:

    (a) PDMS has clogged the needle and is preventing flow. In this case, remove the needle, disconnect it from its tubing, and use a fresh syringe to perfuse sterile PBS through the needle, unclogging it. Once the needle is unclogged, reinsert it into the hole in the TMJ chamber.

(b) Flow has not been properly equalized with the clamps, causing all of the media to flow through one tube. In this case, tighten the clamp on the tube with flow until flow is observed in the other tubes.

(c) Holes in the syringe casing from multiple needle insertion trials are leaking the media rather than directing flow through the needle tubing. Proper needle insertion technique should prevent this from occurring. If it does occur, the assembly may need to be stopped and a new syringe-mold complex built.

References

1. Bancroft GN, Sikavitsas VI, van den Dolder J, Sheffield TL, Ambrose CG, Jansen JA, Mikos AG (2002) Fluid flow increases mineralized matrix deposition in 3D perfusion culture of marrow stromal osteloblasts in a dose-dependent manner. Proc Natl Acad Sci USA 99:12600–12605

2. Marolt D, Marcos Campos I, Bhumiratana S, Koren A, Petridis P, Zhang G, Spitalnik P, Grayson W, Vunjak-Novakovic G (2012) Engineering human bone form embryonic stem cells. Proc Natl Acad Sci USA 109:8705–8709

3. Frohlich M, Grayson WL, Marolt D, Gimble J, Velikonja NK, Vunjak-Novakovic G (2010) Bone grafts engineered from adipose-derived stem cells in perfusion bioreactor culture. Tiss Eng A 16:179–189

4. Bhumiratana S, Grayson WL, Castañeda A, Rockwood DN, Gil ES, Kaplan DL, Vunjak-Novakovic G (2011) Nucleation and growth of mineralized bone matrix on silk-hydroxyapatite composite scaffolds. Biomaterials 32:2812–2820

5. Rockwood DN, Gil ES, Park SH, Kluge JA, Grayson W, Bhumiratana S, Rajkhowa R, Wang X, Kim SJ, Vunjak-Novakovic G, Kaplan DL (2011) Ingrowth of human mesenchymal stem cells into porous silk particle reinforced silk composite scaffolds: An in vitro study. Acta Biomater 7:144–151

6. Sikavitsas VI, Bancroft GN, Holtorf HL, Jansen JA, Mikos AG (2003) Mineralized matrix deposition by marrow stromal osteoblasts in 3D perfusion culture increases with increasing fluid shear forces. Proc Natl Acad Sci USA 100:14683–14688

7. Grayson WL, Marolt D, Bhumiratana S, Fröhlich M, Guo XE, Vunjak-Novakovic G (2011) Optimizing the medium perfusion rate in bone tissue engineering bioreactors. Biotechnol Bioeng 108:1159–1170

8. Ishaug SL, Crane GM, Miller MJ, Yasko AW, Yaszemski MJ, Mikos AG (1997) Bone formation by three-dimensional stromal osteoblast culture in biodegradable polymer scaffolds. J Biomed Mater Res 36:17–28

9. Grayson WL, Fröhlich M, Yeager K, Bhumiratana S, Canizzarro C, Wan LQ, Chan E, Liu X, Guo XE, Vunjak-Novakovic G (2010) Engineering anatomically-shaped human bone grafts. Proc Natl Acad Sci USA 107:3299–3304

Methods in Molecular Biology (2014) 1202: 79–94
DOI 10.1007/7651_2013_31
© Springer Science+Business Media New York 2013
Published online: 7 September 2013

Determining the Role of Matrix Compliance in the Differentiation of Mammary Stem Cells

KangAe Lee and Celeste M. Nelson

Abstract

Multipotent stem cells maintain the structure and function of the mammary gland throughout its development and respond to the physiological demands associated with pregnancy and lactation. The ability of mammary stem cells to maintain themselves as well as to give rise to differentiated progeny is not only affected by soluble factors but has increasingly become linked to mechanical cues including the elastic modulus of the extracellular matrix (ECM). Here we describe a protocol for determining how the mechanical properties of the ECM regulate the fate of mammary stem or progenitor cells. This protocol includes detailed methods for the fabrication of substrata with varying stiffness, culture of mammary progenitor cells on synthetic substrata, pharmacological modulation of actomyosin contractility, and analysis of gene expression to define the resulting fate of human mammary stem cells.

Keywords: Differentiation, Lineage specification, Mammary stem cells, Matrix compliance, Mechanical stress, Mechanotransduction, Myoepithelial cells, Tissue morphogenesis

Abbreviations

αSMA	Alpha-smooth muscle actin
ESA	Epithelial-specific antigen
FAK	Focal adhesion kinase
K	Keratin
MLC	Myosin light chain
MSC	Mesenchymal stem cell
Muc1	Sialomucin-1
ROCK	Rho-associated kinase
TDLU	Terminal ductal lobular unit
2D	Two-dimensional
3D	Three-dimensional

1   Introduction

The mammary gland is a highly dynamic organ that undergoes dramatic morphogenetic changes during puberty, pregnancy, lactation, and post-lactational involution. During pregnancy, the ducts branch laterally to form an expanded tree with concomitant epithelial proliferation and differentiation; upon involution, the secretory epithelium undergoes apoptosis and the gland remodels back essentially to its virgin state (1–3). This regenerative and remodeling capacity of the gland requires multipotent stem or progenitor cells in the mammary epithelium. Mammary stem or progenitor cells provide dynamic and flexible attributes to the mammary gland and give rise to either the mature luminal epithelium or to the myoepithelium through a series of lineage specification and concomitant differentiation (3, 4). Human mammary stem or progenitor cells are characterized by the expression of both keratin (K)14 and K19, and generate the bi-layered terminal ductal lobular unit (TDLU), the basic functional structure of the human mammary epithelium (5, 6). Cells from the two epithelial layers of the TDLU express a number of specific proteins that are frequently used as lineage markers: the inner layer of luminal epithelial cells expresses K8, K18, K19, and sialomucin-1 (Muc1), and the outer layer of myoepithelial cells expresses K5, K14, and α-smooth muscle actin (αSMA) (5–7).

Maintenance and differentiation of mammary progenitor cells are governed by complex interactions between cell–cell, cell-extracellular matrix (ECM), and cell-soluble factors as well as by mechanical cues present within the tissues (2, 7–9). Although the regulatory roles of soluble signals, such as growth factors and cytokines, for stem cells have long been appreciated, recent studies demonstrate that the regulation of stem cell fate by these biochemical signals is also strongly affected by co-existing adhesive, mechanical, and topological cues (8, 10, 11). Moreover, regulation of stem cells in vivo normally occurs in the context of development, tissue regeneration, and wound healing, in which the mechanical environment surrounding the stem cells changes dynamically (10, 12). The modulus of elasticity (E) of the ECM, also commonly but incorrectly referred to as tissue stiffness, not only varies within the body from soft brain tissue (0.1 kPa) to rigid calcifying bone (>30 kPa) (13, 14), but also changes dynamically over the course of development (e.g., mammary branching morphogenesis), in response to function (e.g., mammary gland lactation), and during pathogenesis (e.g., tissue fibrosis and tumorigenesis) (14–18). Such changes in matrix elasticity have been shown to influence cellular behaviors including proliferation, locomotion, spreading, and differentiation of stem cells (13, 17, 19, 20). Culture of multipotent mesenchymal stem cells (MSCs) on synthetic matrix mimicking the elasticity of ECM within a tissue leads to tissue-specific gene expression and promotes organ-specific differentiation (13). Matrix compliance also regulates the renewal and fate-switching of mammary progenitor cells (7).

Stem cells sense and respond to mechanical signals through integrin-mediated adhesions and the force balance between intracellular cytoskeletal contractility and resistance from the ECM (11, 21). Cell-ECM adhesions lead to the phosphorylation of focal adhesion kinase (FAK) and recruitment of vinculin to transmit forces between the ECM and the cytoskeleton, in which myosin-mediated contractility acts as a primary regulator of cellular contractile forces (11, 22, 23). These intracellular forces regulate signaling pathways involved in fundamental cellular processes which play a critical role in determining cell phenotype, such as stem cell differentiation (7, 13, 17). RhoA, a member of the Rho family of small GTPases, regulates signaling involved in cytoskeletal reorganization. RhoA stimulates cytoskeletal tension through its effector, Rho-associated kinase (ROCK), which directly phosphorylates myosin light chain (MLC) and MLC phosphatase to synergistically increase MLC phosphorylation and thus myosin II contractility (24–26). Inhibiting cytoskeletal tension using Y27632 (a ROCK inhibitor), blebbistatin (a myosin II inhibitor), or ML-7 (a MLC kinase inhibitor), or enhancing actomyosin contraction using calyculin A (a MLC phosphatase inhibitor) significantly affects the phenotype of mammary stem cells as well as MSCs (7, 21, 27, 28).

Here, we describe methods for examining how matrix compliance affects the differentiation of human mammary progenitor cells. This protocol presents technical details for the fabrication of synthetic matrices of varying stiffness, culture of mammary progenitor cells on these synthetic substrata, pharmacological modulation of actomyosin contractility, and cellular and molecular characterization of mammary progenitor cell fate in response to alterations in the mechanical properties of the ECM. This series of methods can be used to examine how changing the mechanical properties of the tissue microenvironment affects the phenotypes of stem and progenitor cells as well as differentiated cells (7, 17).

2 Materials

2.1 Cell Culture

1. D920 human mammary progenitor cells (6).

2. H14 medium: 1:1 DMEM:F12 medium supplemented with 250 ng/ml insulin, 10 μg/ml human transferrin, 2.6 ng/ml sodium selenite, 0.1 nM estradiol, 1.4 μM hydrocortisone, 5 μg/ml prolactin, and 0.1 mM gentamicin.

3. Non-pepsinized collagen type I (bovine; Koken).

4. Phosphate-buffered saline (PBS).

5. Trypsin-EDTA (0.05 %).

2.2 Synthesis and Functionalization of Synthetic Substrata

1. Coverglass (31.75 mm diameter, 17–25 mm thick; Fisher Scientific).

2. Germicidal UV lamp (365 nm).

3. Acrylamide (40 % solution).

4. N,N'-Methylene bisacrylamide (2 % solution).

5. Ammonium persulfate (APS).

6. N,N,N',N'-Tetramethylethylenediamine (TEMED).

7. 3-Aminopropyltrimethoxysilane (APTMS; Sigma).

8. Acetone.

9. Glutaraldehyde (50 % stock; Sigma).

10. Dichlorodimethylsilane.

11. Toluene.

12. Methanol.

13. N-Sulfosuccinimidyl-6-(4′-azido-2′-nitrophenylamino) hexanoate (Sulfo-SANPAH; Pierce).

14. 4-(2-Hydroxyethyl)-1-piperazineethanesulfonic acid (HEPES; 50 mM, pH 8.5).

15. Non-pepsinized collagen type I (bovine; Koken).

16. 6-well plate.

2.3 Pharmacological Manipulation of Cell Contractile Machinery

1. Y27632 (Tocris).

2. Blebbistatin (Sigma).

3. Calyculin A (Calbiochem).

2.4 Immunofluorescence Staining and Microscopy Analysis

1. Phosphate-buffered saline (PBS).

2. Paraformaldehyde (4 % in PBS).

3. Triton-X-100 (0.3 % in PBS).

4. Goat serum (10 % in PBS).

5. Rabbit anti-keratin-14 antibody (Covance).

6. Mouse anti-keratin-8 antibody (AbCam).

7. Rabbit anti-phospho-focal adhesion kinase (pFAK^{Tyr397}) antibody (Invitrogen).

8. Rabbit anti-vinculin antibody (Cell Signaling).

9. Alexa Fluor 488 phalloidin (Invitrogen).

10. Alexa 594 goat-anti-rabbit (Invitrogen).

11. Alexa 488 goat-anti-mouse (Invitrogen).

12. Hoechst 33342 (Invitrogen).

13. Inverted fluorescence microscope.

Table 1
Primers used to determine mammary progenitor cell fate via qRT-PCR

Gene	Sequences
18S rRNA	Fwd: CGCCGACGACCCATTCGAAC Rev : GAATCGAACCCTGATTCCCCGTC
Keratin-8	Fwd: AGTTACGGTCAACCAGAG Rev : GTCTCCAGCATCTTGTTC
Keratin-14	Fwd: AACCACGAGGAGGAGATG Rev : GTTCAGAATGCGGCTCAG
Keratin-19	Fwd: GCGACTACAGCCACTACTAC Rev : GTCTCAAACTTGGTTCGGAAG
E-cadherin	Fwd: TGAAGGTGACAGAGCCTCTGGAT Rev : TGGGTGAATTCGGGCTTGTT
P-cadherin	Fwd: GCTGAACATCACGGACAAG Rev : CCTTCCTCGTTGACCTCTG
p63	Fwd: TTGTTGGAAAGTAACTGTGAGAAC Rev : CAAGGGAACTCTTCGTTTAAGTG
αSMA	Fwd: GAGTTACGAGTTGCCTGATG Rev : GGTCCTTCCTGATGTCAATATC

2.5 Quantitative Real-Time PCR (qRT-PCR) Analysis

1. Phosphate-buffered saline (PBS).
2. Trizol reagent (Invitrogen).
3. Diethylpyrocarbonate (DEPC) water.
4. cDNA synthesis kit (Thermo Scientific).
5. iQ SYBR Green Supermix (BioRad).
6. Primers (Table 1).
7. Real-Time PCR detection system (such as MiniOpticon™ Real-Time PCR detection system).
8. UV spectrophotometer.

3 Methods

3.1 Mammary Progenitor Cell Culture

D920 human mammary progenitor cells were derived from epithelial-specific antigen (ESA)$^+$/Muc1$^-$ human breast stem cells and immortalized with human papilloma virus (HPV) proteins E6 and E7 (5). Not only do D920 cells express K19, a marker for mammary stem cells, but clones give rise to cells expressing combinations of K19 and K14 in two-dimensional (2D) culture and generate discretely bi-layered TDLU-like structures in vivo in cleared fat pads of NOD/SCID mice and in three-dimensional

(3D) matrigel (5). The following protocol describes detailed methods to maintain D920 mammary progenitor cells.

1. Coat cell culture plate with collagen (50 µg/ml) and incubate at 4 °C overnight.

2. Rinse the collagen-coated plate twice with 1× PBS.

3. Trypsinize D920 cells using 2 ml of Trypsin-EDTA (0.05 %).

4. Collect cells with 10 ml H14 medium into a 15 ml tube.

5. To remove residual Trypsin-EDTA, centrifuge the suspension at $100 \times g$ for 5 min, aspirate the supernatant, and resuspend cells with fresh H14 medium.

6. Plate D920 human mammary progenitor cells on the collagen-coated plate using H14 medium (see Note 1).

7. Change medium every 48 h.

3.2 Preparation of Synthetic Substrata of Varying Stiffness for Cell Culture

This protocol describes methods to create synthetic matrices with tunable elasticity from polyacrylamide (Fig. 1). Polyacrylamide provides several advantages as a cell culture substratum: its elasticity can be tuned precisely by changing the relative concentrations of acrylamide and bis-acrylamide (20); the surface chemistry of the gel can be kept constant while changing its mechanical properties (17, 20); nonspecific binding of proteins or cells are negligible and thus the adhesive molecules that are covalently attached to the surface serve as the primary ligands for cell attachment (29); the porosity of the gels allows for the flow of medium and provides a more physiological environment than do solid surfaces (30). Table 1 provides the expected elastic moduli for specific concentrations of acrylamide and bis-acrylamide. The viscoelastic properties of the polyacrylamide gels were determined using rheometry, and the Young's moduli of the gels were calculated from the shear modulus, G, using the following equation: $E = 2G(1 + \nu)$, where ν is the Poisson ratio ($\nu = 0.48$ for polyacrylamide) (17, 31). The following protocol is written for 31.75 mm-diameter coverglass for use in a 6-well plate. The volumes can be adjusted up or down to compensate for larger or smaller coverglass, as needed.

1. To render the surface hydrophobic, incubate coverglass in 0.1 N NaOH for 30 min (see Note 2). Rinse coverglass three times thoroughly with MilliQ dH_2O and after the last wash, dry completely using a vacuum aspirator (slides can be stored under nitrogen after this step). Follow steps 2 and 3 for the bottom slide and follow step 4 for the top slide.

2. [Bottom slide] In a chemical fume hood, incubate cover glass in 2 % 3-aminopropyl trimethoxysilane (APTMS) diluted in acetone for 30 min. Rinse each coverglass three times with acetone and air dry (see Note 3; slides can be stored under nitrogen after this step).

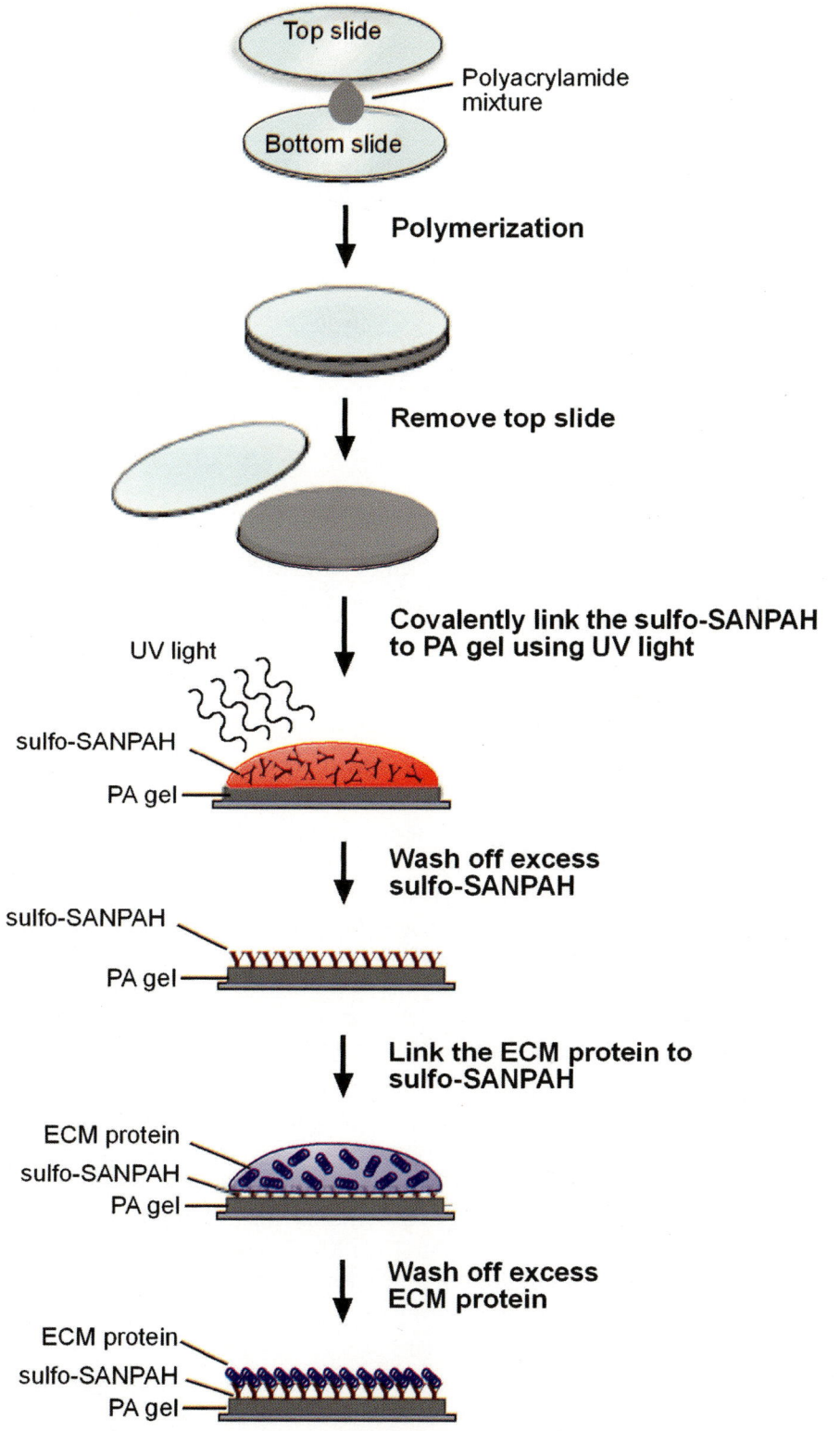

Fig. 1 Schematic overview of the setup for making synthetic substrata. The gel-glass composite contains the amino-silanated coverglass (*bottom slide*), polymerizing acrylamide solution, and chloro-silanated coverglass (*top slide*). The polyacrylamide gel is functionalized by cross-linking sulfo-SANPAH at 365 nm followed by attachment of desired ECM protein

Table 2
Polyacrylamide solutions to generate synthetic substrata of various compliances

Elasticity (Pa)	% Acrylamide	% Bis-acrylamide	Acrylamide (μl)	Bis-acrylamide (μl)	dH$_2$O (μl)	TEMED (μl)	10 % APS (μl)
130	5	0.01	125	5	864.5	0.5	5
910	5	0.03	125	15	854.5	0.5	5
2,030	5	0.06	125	30	839.5	0.5	5
4,020	5	0.35	125	175	694.5	0.5	5

3. [Bottom slide] In a chemical fume hood, incubate coverglass in 0.5 % glutaraldehyde diluted in 1× PBS for 30 min. Rinse coverglass three times with MilliQ dH$_2$O and dry completely with vacuum aspirator. (Slides can be stored under nitrogen after this step.)

4. [Top slide] In a chemical fume hood, incubate coverglass in 2 % dichlorodimethylsilane diluted in toluene for 30 min. Rinse coverglass three times with methanol and air dry. (Slides can be stored under nitrogen after this step.)

5. Prepare polyacrylamide solution using acrylamide and bis-acrylamide at desired final concentration. See Table 2 for concentrations and the corresponding elastic modulus. Combine acrylamide, bis-acrylamide, and MilliQ dH$_2$O in a 1.5 ml tube. Degas the mixture under vacuum for 30 min to remove dissolved oxygen (see Note 4). Just before removing the polyacrylamide solution from the vacuum, prepare 10 % solution of APS. Add appropriate volumes of TEMED and APS to the degassed polyacrylamide solution (Table 2).

6. Place 36 μl of the polyacrylamide mixture onto the center of the bottom slide and cover with the top slide (Fig. 1). The setup resembles a sandwich in which the polymerizing solution sits between the top and bottom coverglass (see Notes 5 and 6).

7. Allow the gel to polymerize for 30 min. After polymerization, place the gels in 1× PBS. These hydrogels can be stored for long periods of time without altering mechanical properties. To store them, immerse hydrogels in 1× PBS to keep them hydrated at 4 °C.

8. Remove the top slide carefully and rinse the gel twice with 1× PBS to remove unpolymerized acrylamide.

The following steps should be performed in a biosafety cabinet using sterilized materials:

9. Incubate the polyacrylamide gel with 100 % ethanol for 5 min and wash three times with 1× PBS. Incubate the gel with 50 mM HEPES, pH 8.5 for 5 min. During the incubation, prepare 2 mM of sulfo-SANPAH dissolved in sterilized MilliQ dH$_2$O (see Note 7).

10. Aspirate excess 50 mM HEPES and place 200 μl of 2 mM sulfo-SANPAH on top of the gel and swirl gently to coat the surface completely (see Note 8).

11. Expose the gel to a germicidal UV lamp (365 nm) for 10 min and rinse once with 50 mM HEPES, pH 8.5.

12. Repeat steps 10–11 and rinse the gel three times with 50 mM HEPES, pH 8.5.

13. Pipet 200 μl of 0.2 mg/ml of collagen diluted in 1× PBS on top of the gel and swirl gently to coat the gel. Incubate gel overnight at 4 °C (see Notes 9–11).

3.3 Culture of Mammary Progenitor Cells on Substrata with Varying Compliance

1. To remove extra collagen from the polyacrylamide surface, wash collagen-coated gels three times with 1× PBS, followed by incubation in culture medium at 37 °C for 1 h.

2. Place gels in a 6-well culture plate.

3. Trypsinize D920 mammary progenitor cells.

4. Plate 1–2 × 10^5 D920 mammary progenitor cells on each gel using H14 medium and culture for 24–72 h.

3.4 Pharmacological Manipulation of Cytoskeletal Contractility

1. Plate D920 mammary progenitor cells on the collagen-functionalized polyacrylamide gel or on a collagen-coated cell culture plate.

2. After 24 h, change medium to fresh H14 medium that includes the desired inhibitor. Y27632 (10 μM) and blebbistatin (12.5 μM) can be used to inhibit cytoskeletal tension. Conversely, calyculin A (2 nM) can be used to enhance actomyosin contraction. A detailed description of these inhibitors is provided above.

3.5 Immunofluorescence Staining to Determine the Fate of Mammary Progenitor Cells

Immunofluorescence staining allows the visualization of specific proteins in cells by binding a specific antibody chemically conjugated with a fluorescent dye such as Alexa fluor. These labeled antibodies directly or indirectly bind to the antigen of interest, which allows for detection of the protein through fluorescence techniques. The fluorescence can be visualized using widefield or confocal microscopy and quantified using a flow cytometer, automated imaging instrument, or imaging software.

1. Rinse cells cultured on collagen-coated gels with 1× PBS and fix in 4 % paraformadehyde solution diluted in 1× PBS for 15 min.

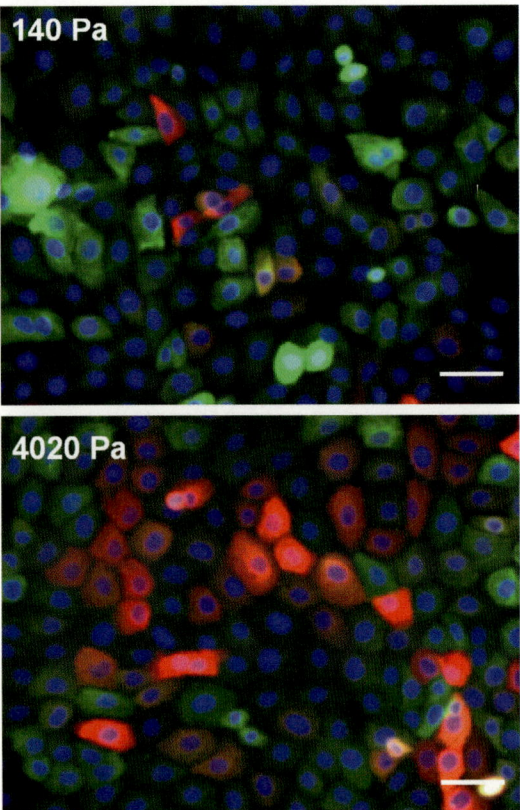

Fig. 2 Immunofluorescence analysis of mammary progenitor cells cultured on substrata of different compliances. Shown are staining for K14 (*red*) and K8 (*green*). Scale bars, 50 μm

2. Rinse samples twice with 1× PBS and permeabilize with 0.3 % Triton-X-100 in 1× PBS for 10 min.

3. Rinse samples twice with 1× PBS and incubate with blocking reagent (0.1 % Triton-X-100 in 10 % goat serum in PBS) for 6 h at room temperature or overnight at 4 °C.

4. Incubate with primary antibody diluted in 10 % goat serum at 4 °C overnight with shaking. Changes in cell fate can be determined using antibodies against cytokeratins (Fig. 2). Alterations in cell-ECM adhesions and cytoskeletal contraction in response to substratum stiffness can be determined using antibodies against FAK or vinculin, or by labeling F-actin with phalloidin. The roles of these proteins are described above.

5. Wash samples three times with 1× PBS for 15 min.

6. Incubate samples with secondary antibody conjugated with fluorescent dye (e.g., Alexa Fluor 488 conjugate) for 2 h at room temperature.

7. Wash samples three times with 1× PBS for 15 min.

8. Counterstain the nuclei of cells with Hoechst 33342 for 10 min prior to imaging.

9. Analyze using ImageJ to determine the number of cells expressing specific keratins.

3.6 Gene Expression Analysis to Determine Mammary Progenitor Fate

Quantitative real-time PCR (qRT-PCR) is a method used to detect relative levels of gene expression. SYBR green is frequently used as a fluorescent dye for qRT-PCR. SYBR intercalates with double-stranded DNA and this intercalation causes the SYBR to fluoresce, which can be detected with a qPCR machine and converted into Ct values from the intensity of the fluorescence. The following protocol covers the qRT-PCR technique using SYBR green methodology and overviews how to analyze data. Lineage specification of mammary progenitor cells can be determined by analyzing the expression levels of luminal epithelial markers (K8, K19, and E-cadherin) or myoepithelial markers (K14, P-cadherin, αSMA, and p63). Sequences for primers to amplify these genes are listed in Table 1.

3.6.1 RNA Isolation from Mammary Progenitor Cells

1. Collect cells by trypsinization and centrifuge at $100 \times g$ for 5 min.

2. Lyse cells in Trizol reagent by repetitive pipetting. Use 1 ml of the reagent for 1–2×10^6 cells. Incubate the homogenized samples for 5 min at room temperature for the complete dissociation of the nucleoprotein complexes.

3. Add 0.2 ml of chloroform per 1 ml Trizol reagent and shake tubes vigorously for 20 s.

4. Centrifuge the sample at $18,000 \times g$ for 15 min. Following centrifugation, the mixture separates into a lower red phase (phenol–chloroform), an interphase, and a colorless upper aqueous phase that contains RNA.

5. Transfer the aqueous phase into a 1.5 ml tube and precipitate the RNA by adding 0.5 ml isopropanol. Mix the sample well by inverting the tube and incubate for 10 min. The RNA precipitate is often invisible at this step.

6. Centrifuge the sample at $18,000 \times g$ at 4 °C for 15 min. The RNA pellet may be visible after centrifugation.

7. Remove the supernatant and wash the RNA pellet twice with 1 ml of 75 % ethanol. Resuspend the RNA pellet by vortexing and centrifuge at $11,000 \times g$ at 4 °C for 5 min.

8. Remove the supernatant (75 % ethanol) and air-dry the RNA pellet (see Note 12).

9. Dissolve RNA in 30–40 μl DEPC-treated water and measure the concentration of RNA by determining the absorbance at 260 nm in a spectrophotometer (see Note 13). Purified RNA can be maintained at −20 or −70 °C for long-term storage (see Note 14).

Table 3
Reverse transcription reaction mix

Reaction mix	Final concentration	Volume (µl)
5× Reaction buffer	1×	4
dNTPs mix (5 mM)	500 µM	2
Oligo dT	500 ng	1
RT enhancer		1
Reverse transcriptase		1
DEPC dH$_2$O		11 − x µl for 1 µg of RNA
RNA	1 µg	x µl for 1 µg of RNA
Total		20

Table 4
Thermal cycles for reverse transcription

Procedure	Temperature (°C)	Time (min)	Number of cycles
cDNA synthesis	42	30	1
Inactivation[a]	95	2	1

[a]After inactivation, place cDNA immediately on ice

3.6.2 cDNA Synthesis Using Reverse Transcription (RT) of RNA

1. Mix 1 µg of RNA with RT-reaction mix as described in Table 3.
2. Spin down reaction mix and follow thermal cycles as described in Table 4.

3.6.3 Quantitative Real-Time PCR (qRT-PCR) Using SYBR Green

1. Mix cDNA with SYBR green reaction mix in 0.2 ml qPCR tube as described in Table 5. Primers used for detecting human mammary progenitor cell fate are listed in Table 1.
2. Spin down reaction mix and run the thermal cycles as described in Table 6 using an optical qPCR thermal cycler.
3. Determine the fold-change in expression of each target mRNA relative to 18S rRNA based on the threshold cycle (Ct) as follows (see Note 15):

$$\text{Fold change} = 2^{-\Delta(\Delta Ct)}[Ct = Ct_{\text{target}} - Ct_{18S}; \ \Delta(\Delta Ct)$$
$$= \Delta Ct_{\text{treatment}} - \Delta Ct_{\text{control}}]$$

Table 5
qRT-PCR reaction mix

Reaction mix	Final concentration	Volume (μl)
2× SYBR Green super mix	1×	12.5
Primer (forward; 5 μM)	100 nM	0.5
Primer (reverse; 5 μM)	100 nM	1
cDNA		1
DEPC dH$_2$O		10
Total		25

Table 6
Thermal cycles for qRT-PCR

Procedure		Temperature	Time	Number of cycles
Polymerase activation and DNA denaturation		95 °C	3 min	1
Amplification	Denaturation	95 °C	10 s	40
	Annealing	55–60 °C	15 s	
	Extension	72 °C	10 s	
	Read plate			
Extension		72 °C	10 min	1
Melt curve analysis		65–98 °C increment 0.2 °C for 0.01 s		1
Read plate Final extension		72 °C	10 min	1

4 Notes

1. The plasticity of mammary progenitor cells may be affected by increasing time and passage in culture. To retain multipotency, D920 cells should be maintained within a narrow passage window and passaged approximately every 5 days when the cells are ~80 % confluent. The multipotency of D920 progenitor cells can be confirmed by expression of K14 and K19 in 2D culture.

2. For uniform gel attachment, coverglass should be coated with an even layer of 0.1 N NaOH. After evaporation, a thin semi-transparent film of NaOH will remain on the coverglass. If needed, repeat this step until a thin white film of NaOH is visible on the surface. Alternatively, this step can be performed by adding 0.1 N NaOH to the surface of the coverglass and heated at 80 °C until the liquid has evaporated.

3. It is important to completely rinse off any residual APTMS, for it will create a brown precipitate with glutaraldehyde that fluoresces under UV light and may thus interfere with later procedures.

4. Dissolved oxygen in the solution will act as a sink for the free radical polymerization. Degassing the solution will not only speed up polymerization but will also ensure more uniform polymerization.

5. Polymerization is initiated immediately after adding 10 % APS and TEMED. To ensure a uniform polymerization of the polyacrylamide gel, step 6 must be performed within a short period of time (couple of minutes).

6. Be careful to avoid air bubbles. The presence of bubbles creates discontinuous regions within the polyacrylamide gel that may cause the mechanical properties to vary spatially within the substratum.

7. Sulfo-SANPAH is light sensitive and should be shielded from light until use.

8. Complete coverage of sulfo-SANPAH is necessary to ensure even coating of ECM proteins on synthetic substrata.

9. Due to the aqueous instability of the sulfosuccinimidyl ester, conjugation of ECM proteins should begin immediately upon activation. If making large gels, place the gel on a rocker to ensure that the solution of ECM protein remains well mixed and that the gels are coated evenly.

10. The concentration of ECM protein can be optimized for each cell type. It is critical that the protein solution not precipitate for proper conjugation of the ECM protein to the polyacylamide gel. Increasing the collagen concentration within the solution to enhance binding may cause the collagen to precipitate once warmed to physiological temperature. We have successfully used concentrations of 0.2 mg/ml of bovine collagen and 0.2 mg/ml of fibronectin, which have previously been shown to support adhesion of D920 mammary progenitor cells (7) and mammary epithelial cells (17).

11. The amount of the ECM protein conjugated to the surface of the polyacrylamide gel can be measured using an enzyme-linked immunosorbent assay (ELISA) or fluorescently labeled

protein (13, 17). Fluorescent staining can also be used to confirm uniform coating of the protein.

12. Do not dry the RNA by centrifugation under vacuum. Overdrying the RNA will greatly decrease its solubility.

13. The ratio of absorbance at 260 nm and 280 nm is used to assess the purity of RNA. A ratio of 260/280 greater than 1.8 is generally accepted as pure for RNA. The ratio of absorbance at 260 and 230 nm is also used as a secondary measure of RNA purity. The 260/230 ratio for pure RNA is often greater than 2.0. Significantly lower ratios may indicate the presence of protein, phenol, or other contaminants that absorb strongly at or near 280 or 230 nm.

14. Freeze and thaw RNA samples as little as possible. It may be better to keep samples on ice for a few hours than to refreeze them in between. Quickly thaw samples then place them on ice.

15. The Ct value is used for accurate quantification of gene expression by qPCR. Ct value should be measured when the level of fluorescence gives signal over the background and is in the linear portion of the amplification curve.

Acknowledgments

We would like to thank Cecillia Lui for technical assistance. This work was supported in part by the NIH (CA128660, GM083997, and HL110335), Susan G. Komen for the Cure (FAS0703855), the David and Lucile Packard Foundation, the Alfred P. Sloan Foundation, and the Camille and Henry Dreyfus Foundation. C.M.N. holds a Career Award at the Scientific Interface from the Burroughs Wellcome Fund.

References

1. Oakes S, Hilton H, Ormandy C (2006) Key stages in mammary gland development—the alveolar switch: coordinating the proliferative cues and cell fate decisions that drive the formation of lobuloalveoli from ductal epithelium. Breast Cancer Res 8(2):207

2. Gjorevski N, Nelson CM (2011) Integrated morphodynamic signalling of the mammary gland. Nat Rev Mol Cell Biol 12(9):581–593

3. Watson C (2006) Key stages in mammary gland development—involution: apoptosis and tissue remodelling that convert the mammary gland from milk factory to a quiescent organ. Breast Cancer Res 8(2):203

4. Watson C, Khaled WT (2008) Mammary development in the embryo and adult: a journey of morphogenesis and commitment. Development 135:995–1003

5. Gudjonsson T, Villadsen R, Nielsen HL, Rønnov-Jessen L, Bissell MJ, Petersen OW (2002) Isolation, immortalization, and characterization of a human breast epithelial cell line with stem cell properties. Genes Dev 16(6):693–706

6. Villadsen R, Fridriksdottir AJ, Rønnov-Jessen L, Gudjonsson T, Rank F, LaBarge MA, Bissell MJ, Petersen OW (2007) Evidence for a stem cell hierarchy in the adult human breast. J Cell Biol 177(1):87–101

7. Lui C, Lee K, Nelson CM (2012) Matrix compliance and RhoA direct the differentiation of mammary progenitor cells. Biomech Model Mechanobiol 11:1241–1249

8. LaBarge MA, Petersen OW, Bissell MJ (2007) Of microenvironments and mammary stem cells. Stem Cell Rev 3(2):137–146

9. Nelson CM, Bissell MJ (2006) Of extracellular matrix, scaffolds, and signaling: tissue architecture regulates development, homeostasis, and cancer. Annu Rev Cell Dev Biol 22 (1):287–309

10. Schedin P, Keely PJ (2011) Mammary gland ECM remodeling, stiffness, and mechanosignaling in normal development and tumor progression. Cold Spring Harb Perspect Biol 3(1):doi:10.1101/cshperspect.a003228

11. Sun Y, Chen CS, Fu J (2012) Forcing stem cells to behave: a biophysical perspective of the cellular microenvironment. Annu Rev Biophys 41:519–542

12. Reilly GC, Engler AJ (2010) Intrinsic extracellular matrix properties regulate stem cell differentiation. J Biomech 43(1):55–62

13. Engler AJ, Sen S, Sweeney HL, Discher DE (2006) Matrix elasticity directs stem cell lineage specification. Cell 126(4):677–689

14. Butcher DT, Alliston T, Weaver VM (2009) A tense situation: forcing tumour progression. Nat Rev Cancer 9(2):108–122

15. Gjorevski N, Nelson CM (2012) Mapping of mechanical strains and stresses around quiescent engineered three-dimensional epithelial tissues. Biophys J 103(1):152–162

16. Paszek MJ, Weaver VM (2004) The tension mounts: mechanics meets morphogenesis and malignancy. J Mammary Gland Biol Neoplasia 9(4):325–342

17. Lee K, Chen QK, Lui C, Cichon MA, Radisky DC, Nelson CM (2012) Matrix compliance regulates Rac1b localization, NADPH oxidase assembly, and epithelial-mesenchymal transition. Mol Biol Cell 23:4097–4108

18. Lopez JI, Kang I, You W-K, McDonald DM, Weaver VM (2011) In situ force mapping of mammary gland transformation. Integr Biol 3 (9):910–921

19. Schrader J, Gordon-Walker TT, Aucott RL, van Deemter M, Quaas A, Walsh S, Benten D, Forbes SJ, Wells RG, Iredale JP (2011) Matrix stiffness modulates proliferation, chemotherapeutic response, and dormancy in hepatocellular carcinoma cells. Hepatology 53 (4):1192–1205

20. Pelham RJ, Wang Y (1997) Cell locomotion and focal adhesions are regulated by substrate flexibility. Proc Natl Acad Sci U S A 94 (25):13661–13665

21. McBeath R, Pirone DM, Nelson CM, Bhadriraju K, Chen CS (2004) Cell shape, cytoskeletal tension, and RhoA regulate stem cell lineage commitment. Dev Cell 6(4):483–495

22. Hoffman BD, Grashoff C, Schwartz MA (2011) Dynamic molecular processes mediate cellular mechanotransduction. Nature 475 (7356):316–323

23. Shi Q, Boettiger D (2003) A novel mode for integrin-mediated signaling: tethering is required for phosphorylation of FAK Y397. Mol Biol Cell 14(10):4306–4315

24. Amano M, Ito M, Kimura K, Fukata Y, Chihara K, Nakano T, Matsuura Y, Kaibuchi K (1996) Phosphorylation and activation of myosin by Rho-ASSOCIATED Kinase (Rho-kinase). J Biol Chem 271(34):20246–20249

25. Ishizaki T, Naito M, Fujisawa K, Maekawa M, Watanabe N, Saito Y, Narumiya S (1997) p160ROCK, a Rho-associated coiled-coil forming protein kinase, works downstream of Rho and induces focal adhesions. FEBS Lett 404(2–3):118–124

26. Kimura K, Ito M, Amano M, Chihara K, Fukata Y, Nakafuku M, Yamamori B, Feng J, Nakano T, Okawa K, Iwamatsu A, Kaibuchi K (1996) Regulation of myosin phosphatase by Rho and Rho-associated kinase (Rho-kinase). Science 273(5272):245–248

27. Kilian KA, Bugarija B, Lahn BT, Mrksich M (2010) Geometric cues for directing the differentiation of mesenchymal stem cells. Proc Natl Acad Sci U S A 107(11):4872–4877

28. Song W, Lu H, Kawazoe N, Chen G (2011) Adipogenic differentiation of individual mesenchymal stem cell on different geometric micropatterns. Langmuir 27(10):6155–6162

29. Kandow CE, Georges PC, Janmey PA, Beningo KA (2007) Polyacrylamide hydrogels for cell mechanics: steps toward optimization and alternative uses. Methods Cell Biol 83:29–46

30. Wang Y, Pelham RJ (1998) Preparation of a flexible, porous polyacrylamide substrate for mechanical studies of cultured cells. Methods Enzymol 298:489–496

31. Boudou T, Ohayon J, Picart C, Tracqui P (2006) An extended relationship for the characterization of Young's modulus and Poisson's ratio of tunable polyacrylamide gels. Biorheology 43(6):721–728

Methods in Molecular Biology (2014) 1202: 95–102
DOI 10.1007/7651_2013_32
© Springer Science+Business Media New York 2013
Published online: 7 September 2013

Conjugation of Proteins to Polymer Chains to Create Multivalent Molecules

Anthony Conway, Dawn P. Spelke, and David V. Schaffer

Abstract

The activation of cellular signaling cascades, critical for regulating cell function and fate, often involves changes in the organization of receptors in the cell membrane. Using *synthetic* multivalent ligands to control the nanoscale organization of cellular receptors into clusters is an attractive approach to elicit desired downstream cellular responses, since multivalent ligands can be significantly more potent than their corresponding monovalent ligands. Synthetic multivalent ligands can serve as both versatile biological tools and potent nanoscale therapeutics, for example in applications to harness them to control stem cell fate in vitro and in vivo. Here we describe the use of recombinant protein expression and bioconjugate chemistry to synthesize multivalent ligands that have the potential to regulate cell signaling in a variety of cell types.

Keywords: Multivalency, Stem cells, Conjugation, Biopolymers

1 Introduction

Multivalent ligand interactions, characterized by multiple copies of a ligand in close proximity simultaneously interacting with multiple binding partners, are ubiquitous in cellular signal transduction (1, 2). Membrane-bound ligand and receptor pairs—such as Delta/Jagged and Notch (3), ephrins and Ephs (4), and receptors in the immune system—are involved in juxtacrine signaling between neighboring cells and often form multimeric assemblies upon binding. Also, some soluble signaling molecules can oligomerize due to posttranslational modifications, including Sonic hedgehog (Shh) (5). Additionally, many growth factors—such as fibroblast growth factors (FGF) (6) and vascular endothelial growth factor (VEGF) (7)—bind heparin and/or extracellular matrix (ECM) proteins and may thus be tethered in close proximity upon secretion from a cell. Such multivalent ligand–receptor interactions are sometimes required for downstream signaling initiation (8), and the degree of multivalency may dictate the potency of numerous elicited responses (9). Therefore, the generation of multivalent versions of synthetic or recombinant forms of these ligands may aid in both basic investigation of signaling mechanisms and the development of high potency signaling molecules for therapeutic application.

The current established method for oligomerizing proteins involves antibody-induced clustering (8); however, this method is not well controlled, resulting in a range of oligomer sizes which can undergo dissociation. Here, we describe a protocol for synthesizing multivalent ligands in a modular fashion via chemical conjugation of recombinant protein to long hyaluronic acid (HyA) polymer chains. We have successfully utilized this technique to develop multivalent Sonic hedgehog (Shh) and demonstrate that higher valency Shh increases potency (10). HyA polymers are flexible, allowing for rotation of bound ligands, and by varying the conjugation ratio of protein to HyA a range of valencies can be achieved. These multivalent ligands can be employed to activate pathways that benefit from receptor clustering, study the basic role of ligand/receptor clustering in signaling, and generate potent signaling agonists for therapeutic application.

2 Materials

2.1 Recombinant Protein Production and Purification

2.1.1 Recombinant Protein Production

1. LB media: 1 % (w/v) tryptone peptone, 0.5 % (w/v) yeast extract, and 0.5 % (w/v) NaCl in water. Adjust pH to 7. Autoclave and store at room temperature.

2. LB agar plate with antibiotic: Add agar at 1.5 % (w/v) to LB media. Autoclave and allow to cool to ~50 °C (able to hold in hands). Add antibiotic at appropriate concentration (e.g., 0.1 mg/mL ampicillin). Pour plates using sterile technique and store at 4 °C.

3. TB media: 1.2 % (w/v) tryptone peptone, 2.4 % (w/v) yeast extract, and 0.4 % glycerol in water. Adjust pH to 7. Autoclave and store at room temperature.

4. L-arabinose (Sigma-Aldrich) or alternative protein induction reagent.

5. Lysis buffer: 5 mM imidazole (Sigma-Aldrich), 50 mM KH_2PO_4, and 300 mM KCl in water. Adjust pH to 8. Filter sterilize, and store at 4 °C. Degas before use.

6. Phenylmethanesulfonyl fluoride (PMSF, Sigma-Aldrich): Make 10 mg/mL stock in isopropanol. Store at 4 °C. Note: PMSF is highly toxic, so follow correct safety protocols when handling.

7. β-Mercaptoethanol (β-ME).

8. Lysozyme, from chicken egg white (Sigma-Aldrich).

2.1.2 Protein Purification	1. Purification materials, e.g., columns, affinity media, and wash and elution buffers.

1. Purification materials, e.g., columns, affinity media, and wash and elution buffers.

2. Dialysis tubing of appropriate molecular weight cut off (MWCO) for protein and dialysis accessories (Spectrum Labs).

3. Dialysis buffer 1: 2 mM EDTA and 150 mM NaCl in PBS. Adjust pH to 6.5. Filter sterilize and store at room temperature. Degas before use.

4. Dialysis buffer 2: 2 mM EDTA, 10 % glycerol in PBS. Adjust pH to 6.5. Filter sterilize and store at room temperature. Degas before use.

5. BCA Protein Assay (Pierce).

6. SDS-PAGE materials.

2.2 Bioconjugation and Multivalent Molecule Characterization

2.2.1 Bioconjugation

1. MES Buffer: 0.1 M 2-(*N*-morpholino)ethanesulfonic acid (MES, Sigma-Aldrich) in water. Adjust pH to 6.5. Store at room temperature. Degas before use.

2. Sodium hyaluronic acid (HyA), MW ~800 kDa (Genzyme, Lifecore, etc.).

3. 3,3′-*N*-(e-Maleimidocaproic acid) hydrazide, trifluoroacetic acid salt (EMCH, Pierce), 1.2 mg/mL working solution in MES buffer.

4. *N*-Hydroxysulfosuccinimide (Sulfo-NHS, Pierce), 2.8 mg/mL working solution in MES buffer.

5. 1-Ethyl-3-(3-dimethylaminopropyl) carbodiimide hydrochloride (EDC, Pierce), 10 mg/mL working solution in MES buffer.

6. Dialysis tubing of appropriate molecular weight cut off (MWCO) for HyA (e.g., 100 kDa MWCO for 800 kDa HyA) and dialysis accessories (Spectrum Labs).

7. Dialysis buffer 3: 2 mM EDTA, 10 % glycerol in PBS. Adjust pH to 7. Filter sterilize and store at room temperature. Degas before use.

8. Tris(2-carboxyethyl)phosphine hydrochloride (TCEP-HCl, Pierce).

9. Dialysis buffer 4: 2 mM EDTA in PBS. Adjust pH to 7. Filter sterilize and store at room temperature. Degas before use.

10. Sodium azide.

11. Penicillin Streptomycin (Life Technologies).

12. Spectra Gel Absorbent (Spectrum Labs).

3   Methods

3.1   Recombinant Protein Production and Purification

To create multivalent bioconjugates, one must first design and produce recombinant proteins that present functional groups that can be used for tethering to polymer chains. Care should be taken determining the appropriate site to attach a chemically labile functional group to assure that the bioactivity of the protein is maintained. In this protocol, we will be using a protein that has been recombinantly modified to display a cysteine at its C-terminus. The protein must also be purified, and a protein purification tag can facilitate this process. We commonly use a hexahistidine tag and purify using immobilized metal (such as Ni^{2+}) affinity chromatography (IMAC). The modified protein must be subcloned into a protein expression vector that is amenable to producing large amount of protein (e.g., pBAD, Life Technologies). Finally, transformation of this plasmid into a bacterial cell strain optimized for protein production (e.g., BL21, Life Technologies) should be performed.

3.1.1   Recombinant Protein Production

1. Using a frozen aliquot of BL21 *E. coli* transformed (11) with pBAD containing the recombinant protein, streak a bacterial LB agar plate with the appropriate antibiotic (e.g., 0.1 mg/mL ampicillin for BL21 *E. coli*) to ensure that only transformed colonies grow (see Note 1). Culture overnight (14–16 h) at 37 °C.

2. Pick one colony and inoculate 100 mL of liquid LB media with appropriate antibiotic added. Culture overnight (16 h) shaking at 37 °C.

3. Inoculate 1 L of liquid TB media, containing the appropriate antibiotic, with 25 mL of overnight LB culture, and split into two aliquots of 500 mL in Erlenmeyer flasks of at least 1 L in volume each. Shake at 37 °C until the OD_{600} of the culture media is ~0.6 (2–4 h).

4. Add appropriate protein induction reagent to the appropriate concentration (e.g., 0.1 % (w/v) L-arabinose for the pBAD vector). Shake at 30 °C for 5 h.

5. Pellet cells in 250 mL polypropylene bottles, 5,000 × *g* for 20 min at 4 °C.

6. Pour off supernatant and freeze cell pellet at −80 °C until use (see Note 2).

From this point on, all steps should be performed at 4 °C unless otherwise specified.

7. Add PMSF stock to cold lysis buffer to a final concentration of 200 µg/mL, β-ME to 20 mM, and lysozyme to 1 mg/mL.

Thaw cells in lysis buffer (30 mL for 1 L culture) by swirling buffer constantly on ice (see Note 3).

8. Incubate at 4 °C for 30 min.

9. Sonicate cell suspension using sonicator (Sonicator 3000, Giltron) on power setting 7. For 30 mL cell suspension in 50 mL Falcon tubes, the power output reading on the sonicator should read ~57 W. Pulse 10 s on, 10 s off, for a total of 240 s.

10. Centrifuge sonicated suspension at 28,000 × g, 60 min, 4 °C.

11. Carefully pipet or decant supernatant into a 50 mL Falcon tube (see Note 4).

3.1.2 Protein Purification

1. Purify supernatant using IMAC chromatography or an analogous method (12). Add β-ME to appropriate solutions directly before purification to reduce disulfide bond formation (see Note 5).

2. Determine concentration of protein of interest on a NanoDrop spectrophotometer (Thermo Scientific) at OD_{280} using the protein's extinction coefficient if known. Bovine serum albumin (BSA) is often used as a standard if the extinction coefficient of the protein of interest is unavailable.

3. Pool protein-containing fractions into dialysis membranes of the appropriate molecular weight cut off (MWCO) size and dialyze at 4 °C in 1,000 mL dialysis buffer 1 for 4 h while constantly stirring (see Note 6).

4. Discard dialysis buffer 1, then add 1,000 mL of dialysis buffer 2, and dialyze at 4 °C overnight while stirring (see Note 7).

5. Validate protein concentration using a BCA assay and protein purity by running an SDS-PAGE gel (13).

3.2 Bioconjugation and Multivalent Molecule Characterization

Protein bioconjugation is a useful technique for a variety of applications. For instance, it can allow for display of biological ligands in a tethered orientation from a material for prolonged signaling compared to soluble ligands. Conjugation of ligands to long polymer chains also allows for multivalent display of ligands, which often have more potent signaling properties compared to monovalent or divalent ligands. In the following protocol, we will discuss how to achieve such multivalent molecules and their subsequent characterization.

3.2.1 Bioconjugation

1. Prepare a 3 mg/mL solution of HyA in degassed MES buffer by stirring very slowly at 4 °C for at least 4 h (see Note 8).

2. Equilibrate EMCH, Sulfo-NHS, and EDC stocks to room temperature.

3. Make working solution (refer to Section 2.2.1, steps 3–5) containing EMCH, Sulfo-NHS, and EDC (see Note 9).

4. Add working solution to fully dissolved HyA solution. Allow to react for 4 h while stirring slowly at 4 °C to create activated HyA-EMCH.

5. Remove the reactants from the product by dialyzing using appropriate MWCO dialysis tubing in 1,000 mL dialysis buffer 3 for 4 h at 4 °C while stirring (see Note 10).

6. Discard dialysis buffer and repeat step 5 two more times (see Note 11).

7. Add TCEP-HCl to thawed purified recombinant protein in 200-fold molar excess, and allow to react for 5 min at 4 °C while stirring.

8. Add activated HyA-EMCH solution in desired molar conjugation ratios (10), and bring all reactions to equal volumes using dialysis buffer 3 (see Note 12).

9. Purge reaction vial headspace with nitrogen, cover vial with aluminum foil to reduce light exposure, and react at 4 °C overnight.

10. Dialyze with dialysis buffer 4 for 4 h at 4 °C while stirring (see Note 13).

11. Discard dialysis buffer and repeat step 11 two more times (see Note 10).

12. If necessary, reduce reaction volume using Spectra Gel Absorbent after the last dialysis step (see Note 14).

3.2.2 Multivalent Molecule Characterization

1. Quantify protein concentration of purified conjugates using a BCA assay.

2. Characterize true conjugation ratios using size-exclusion chromatography coupled with multi-angle light scattering (SEC-MALS) (14) as described (15) on an HPLC with the appropriate column (e.g., PolySep-GFC-P 6000, Phenomenex).

4 Notes

1. Make sure to streak a new LB agar plate before every protein production, since using old plates may cause issues with protein folding and solubility during bacterial growth.

2. Freezing the cell pellet can help disrupt the cell membrane to increase protein yields.

3. Make sure to avoid excess bubbles during pellet resuspension in lysis buffer, as the surface tension from bubbles tends to denature proteins.

4. Avoid saving hazy supernatant, as this is indicative of insoluble proteins and lipids that may overload the IMAC column and thus decrease the purity of the final protein sample, while increasing the overall amount of protein in the final eluate.

5. Save 5–10 mL of dialysis buffer 2 for BCA assay and ~1 mL elution buffer for spectrophotometry measurements.

6. For most proteins, 3–5 kDa MWCO dialysis tubing is recommended.

7. Glycerol in the dialysis buffer will concentrate the protein sample as well as increase stability of the protein in solution.

8. Stirring slowly with a stir bar is necessary to fully solubilize high molecular weight HyA. Too vigorous stirring will, however, physically shear the long polymer chains and result in decreased average molecular weight, which is to be avoided.

9. It is advised to dissolve these reagents in a minimal volume of MES buffer to increase their concentration during the reaction with HyA. Sterile filter the working solution after fully dissolved.

10. For 800 kDa HyA, it is recommended to use 100 kDa MWCO tubing, to remove any lower molecular weight impurities.

11. Save approximately 50 mL of fresh dialysis buffer 3 for use in the subsequent conjugation reactions.

12. Conjugation reaction volumes should be 5–10 mL.

13. To sterilize conjugates, add 0.02 % (w/v) sodium azide and 1 % (v/v) Pen-Strep to the first dialysis volume.

14. Using aluminum foil as a wrapper, pack absorbent gel around the dialysis tubing, and place upright in cold room *for no more than 2 h*. Extended incubation will precipitate samples and render them inactive. Physically remove absorbent powder from exterior of dialysis tubing, followed by dipping the tubing into cold sterile water to completely remove powder.

Acknowledgments

This work was supported by the California Institute for Regenerative Medicine (CIRM) grant RT2-02022. A.C. was partially supported by a training grant fellowship from CIRM (T1-00007). D.P.S. was partially supported by an NSF Graduate Research Fellowship and a training grant fellowship from CIRM (TG2-01164).

References

1. Mammen M, Choi S-K, Whitesides GM (1998) Polyvalent interactions in biological systems: implications for design and use of multivalent ligands and inhibitors. Angew Chem Int Ed 37:2754–2794

2. Kiessling LL, Gestwicki JE, Strong LE (2000) Synthetic multivalent ligands in the exploration of cell-surface interactions. Curr Opin Chem Biol 4:696–703

3. Artavanis-Tsakonas S, Rand MD, Lake RJ (1999) Notch signaling: cell fate control and signal integration in development. Science 284:770–776

4. Ashton RS, Conway A, Pangarkar C, Bergen J, Lim KI, Shah P, Bissell M, Schaffer DV (2012) Astrocytes regulate adult hippocampal neurogenesis through ephrin-B signaling. Nat Neurosci 15:1399–1406

5. Vyas N, Goswami D, Manonmani A, Sharma P, Ranganath HA, VijayRaghavan K, Shashidhara LS, Sowdhamini R, Mayor S (2008) Nanoscale organization of hedgehog is essential for long-range signaling. Cell 133:1214–1227

6. Ye S, Luo Y, Lu W, Jones RB, Linhardt RJ, Capila I, Toida T, Kan M, Pelletier H, McKeehan WL (2001) Structural basis for interaction of FGF-1, FGF-2, and FGF-7 with different heparan sulfate motifs. Biochemistry 40:14429–14439

7. Krilleke D, Ng YS, Shima DT (2009) The heparin-binding domain confers diverse functions of VEGF-A in development and disease: a structure-function study. Biochem Soc Trans 37:1201–1206

8. Davis S, Gale NW, Aldrich TH, Maisonpierre PC, Lhotak V, Pawson T, Goldfarb M, Yancopoulos GD (1994) Ligands for EPH-related receptor tyrosine kinases that require membrane attachment or clustering for activity. Science 266:816–819

9. Arlaud GJ, Colomb MG, Gagnon J (1987) A functional-model of the human C1 complex—emergence of a functional-model. Immunol Today 8:106–111

10. Wall ST, Saha K, Ashton RS, Kam KR, Schaffer DV, Healy KE (2008) Multivalency of Sonic hedgehog conjugated to linear polymer chains modulates protein potency. Bioconjug Chem 19:806–812

11. Maniatis T, Fritsch EF, Sambrook J (1982) Molecular cloning : a laboratory manual. Cold Spring Harbor Laboratory, Cold Spring Harbor, NY

12. Zachariou M (2008) Affinity chromatography: methods and protocols. Humana, Totowa, NJ

13. Hames BD (1998) Gel electrophoresis of proteins : a practical approach. Oxford University Press, Oxford

14. Yu Y, DesLauriers PJ, Rohlfing DC (2005) SEC-MALS method for the determination of long-chain branching and long-chain branching distribution in polyethylene. Polymer 46:5165–5182

15. Pollock JF, Ashton RS, Rode NA, Schaffer DV, Healy KE (2012) Molecular characterization of multivalent bioconjugates by size-exclusion chromatography with multiangle laser light scattering. Bioconjug Chem 23:1794–1801

Methods in Molecular Biology (2014) 1202: 103–110
DOI 10.1007/7651_2013_37
© Springer Science+Business Media New York 2013
Published online: 24 October 2013

An Assay to Quantify Chemotactic Properties of Degradation Products from Extracellular Matrix

Brian M. Sicari, Li Zhang, Ricardo Londono, and Stephen F. Badylak

Abstract

The endogenous chemotaxis of cells toward sites of tissue injury and/or biomaterial implantation is an important component of the host response. Implanted biomaterials capable of recruiting host stem/progenitor cells to a site of interest may obviate challenges associated with cell transplantation. An assay for the identification and quantification of chemotaxis induced by surgically placed biologic scaffolds composed of extracellular matrix is described herein.

Keywords: Chemotaxis, Stem cells, Cryptic peptides, Extracellular matrix

1 Introduction

One of the purported methods by which biologic scaffold materials composed of extracellular matrix (ECM) support constructive tissue remodeling involves the endogenous recruitment of stem and progenitor cells to the site of interest (1–10). This cell recruitment process has been attributed, in part, to chemotactic cryptic peptides generated during the process of in vivo scaffold degradation (11, 12). Specific peptides have been described with such chemotactic activity (13–15).

The directed movement of cells toward sites of tissue injury, inflammation, and infection plays a crucial role in homeostasis and response to injury (16–19). Identification of the bioactive factors that mediate such processes (i.e., specific stimuli) is important for the understanding of disease processes, understanding host physiologic and pathologic responses, and for the development of innovative and effective therapeutic strategies. This manuscript describes an in vitro assay to evaluate the chemotactic response of perivascular stem cells (PVSC) (3–6). These cells were originally isolated from human skeletal muscle and have been suggested to be a critical component of the mammalian response to injury in several anatomic locations (3, 20–24). The in vitro chemotactic assay described herein can be used for any cell type provided that the individual parameters of the assay are optimized for the cell type of interest.

Considering the current interest in various forms of cell therapy, the challenges in maintaining the presence and viability of cells at tissue sites of interest, and the potential role of paracrine factors in recruiting various cells to specific anatomic locations, this in vitro assay can be a valuable laboratory tool.

2 Materials

1. Cell type of interest (See Note 1).
2. 48 Well Micro Chemotaxis Chamber (Neuro Probe, Gaithersburg, MD, USA) (Fig. 1a).
3. Polycarbonate membrane filter (Neuro Probe, Gaithersburg, MD, USA) with 8 μm pore size (See Note 2).
4. Accessory pack for polycarbonate filters (Neuro Probe, Gaithersburg, MD, USA).
5. Collagen I from rat tail (BD Biosciences, Franklin Lakes, NJ, USA) (See Note 3).
6. 95 % methanol.
7. Diff-Quik Stain Set (Source: Dade Behring, Deerfield, IL, USA, or equivalent) or a fluorescent nuclear stain such as Draq5 (Cell Signaling, Danvers, MA, USA) or 4',6-diamidino-2-phenylindole (DAPI; BD Biosciences, Franklin Lakes, NJ, USA).
8. 2″ × 3″ Glass slides (Source: Kimble Scientific, Vineland, NJ, USA, or equivalent).
9. Cover slip.

3 Methods

3.1 Overview

Potential chemotactic substances to be evaluated are placed in the lower wells of the chemotaxis chamber. Migrating cells are placed in the upper wells. A porous filter membrane separates the upper and lower wells. As the substances diffuse cells begin to migrate and are trapped in the membrane where they can subsequently be stained and quantified (Fig. 1b).

3.2 Filter Preparation

1. Choose a polycarbonate filter with appropriate pore size (see Note 2). The appropriate pore size for PVSCs is 8 μm.
2. Coat filter with appropriate adhesion-promoting molecule (see Note 3). Collagen I is used as a substrate for PVSC chemotaxis.

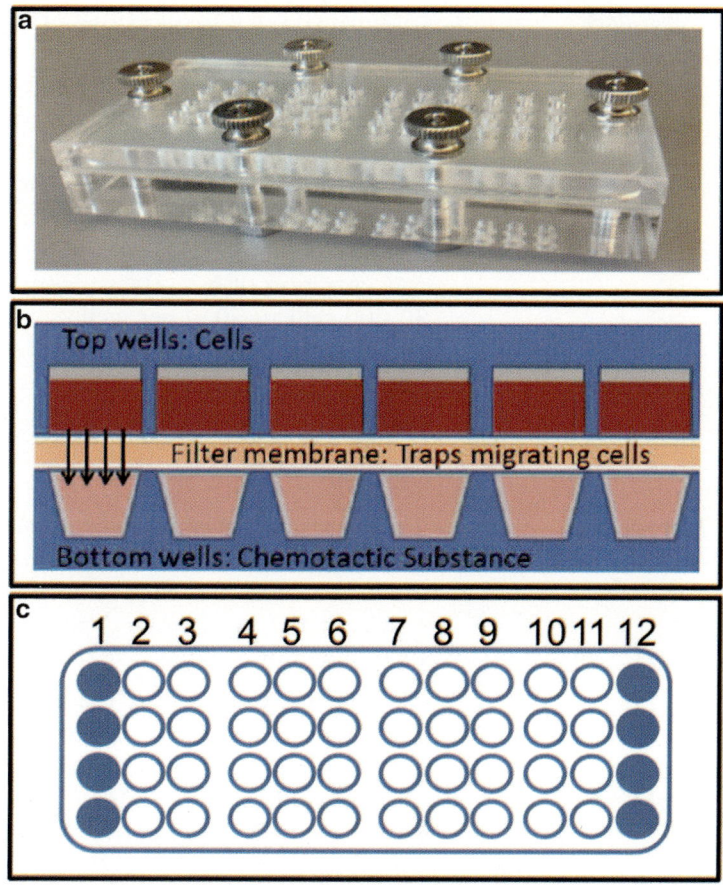

Fig. 1 An established assay for the identification of chemotaxis. Chemotaxis chamber used for the cell migration assay (**a**). Schematic representation of the chemotaxis assay (**b**). Outermost wells (*solid circles*) of the chamber are not used due to artifacts from edge evaporation (**c**). In a 48-well chamber up to ten substances can be tested (*columns 2* through *11*) in quadruplicate

(a) Pour 0.05 mg/ml collagen I solution (prepared according to the manufacturer's protocol) into a large 100 mm cell culture dish.

(b) Immerse filters in collagen I solution, matte side down, making sure that both sides are coated.

(c) Incubate filters in collagen solution for 30–45 min at room temperature.

(d) Float filters over PBS, first with one side down in the PBS and then with the other side down in the PBS.

(e) Clip large filter clamp (from Neuro Probe accessory pack) on the edge of one end of the filter, and hang at room temperature for 20 min to dry. Use filter for assay within 4 h of removal from the collagen I solution.

3.3 Preparation of Responding Cells

1. Established cell lines or primary cells being expanded in culture are serum starved for 18–24 h prior to the chemotaxis assay in basal growth media with 0.5 % serum. On the day of the assay, the cultured cells are trypsinized, counted, and resuspended in basal growth media without serum (migration media) to achieve desired cell density for the assay (see Note 4). Freshly isolated cells are counted and resuspended in migration media at desired cell density (see Note 4). PVSCs are resuspended at a concentration of 6×10^5 cells/mL, to allow easy pipetting of 30,000 cells in 50 µl per well.

2. Place cell suspension in a conical tube with loosened cap, and incubate at 5 % CO_2/37 °C for 1 h. The chemoattractants and chemotaxis chamber are prepared during this incubation period.

3.4 Preparation of Chemoattractants

1. The chemoattractants and control chemoattractants are prepared and diluted in migration media at desired concentrations (see Note 5). Migration media alone and 20 % serum represent the negative and positive controls, respectively, for PVSC chemotaxis.

2. The chemoattractants are loaded into the bottom plate of the chemotaxis chamber on a flat bench top outside of the cell culture hood (see Note 6). Set up assay with quadruplicates when possible, and exclude edge wells due to potential artifacts caused by media evaporation (Fig. 1c).

3. Adjust micropipette so that filled wells on the bottom plate of the chemotaxis chamber form a slight positive meniscus. This volume is 27.5 µl for most liquids and 28 µl for more viscous samples.

3.5 Preparation of Chemotaxis Chamber

1. Carefully place filter membrane (matte side down) evenly on top of the filled wells of the bottom plate of the chemotaxis chamber without leaving any bubbles underneath the filter membrane (see Note 7).

2. Carefully add silicone gasket.

3. Push top plate down against bottom plate, and hold it firmly while applying thumbnuts (see Note 8).

4. Remove cells from incubator, and gently pipet cells to resuspend evenly.

5. Pipet 50 µl of resuspended cells into each upper well of the chemotaxis chamber (see Note 9).

6. Incubate at 5 % CO_2/37 °C for 1 to 18 h, depending on cell type (see Note 10). Three hours is the optimal incubation period for PVSCs.

3.6 Disassembly of Chemotaxis Chamber

1. While firmly holding down the top plate, remove thumbnuts. Invert the entire chamber onto a paper towel while grasping the corners of the top plate (now on the bottom), and slowly lower the top plate until it rests on the bench top, making sure that the filter stays stuck to the gasket and top plate. The filter is now on top of the gasket, and migrated cells are facing up on the filter. The side of the filter now facing up is referred to as the migrated cell side (see Note 11).

    (a) Carefully (see Note 12) and quickly lift just the end of the filter, and clamp 1 mm of the edge with the large plastic clamp (from Neuro Probe accessory pack). Take care to place the clamp evenly on the filter, with the filter centered in the jaws of the clamp and the clamp parallel to the ends of the filter. Lift the filter with the plastic clamp, and while lifting the filter immediately attach the small metal clamp (from Neuro Probe accessory pack) to the other end in an even fashion so that the filter is centered in the jaws of the small clamp. Take care so that the filter does not fold over on itself.

    (b) Holding both clamps, with the migrated cell side up, wet the underside of the filter in a dish containing PBS. Take care not to allow the PBS to wash over the cell side of the filter.

    (c) While holding the filter by the large clamp, with the small clamp attached to the other end and hanging free, wipe the cells off the non-migrated side of the filter by drawing the filter up over the wiper blade (from Neuro Probe accessory pack). The blade should first contact the filter just below the jaws of the wide clamp. Apply light pressure evenly against the blade, and maintain an angle of about 30° from vertical for the portion of the filter above the wiper. It is important to complete the wiping carefully and quickly so that the cells do not dry on the filter; drying occurs in 10–20 s and will prevent complete removal of the non-migrated cells.

    (d) Clean the wiper with a kimwipe.

    (e) Repeat steps (b)–(d) two to three times.

3.7 Filter Fixation and Analysis

1. To fix the filter, carefully float on 95 % methanol in a 100 mm petri dish, migrated cell side down. Then carefully and evenly submerge the filter for 2 min.

2. Allow the filter to dry with migrated cell side up.

3. Using the small metal clamps placed evenly on both sides of the filter, stain as follows (see Note 13):

(a) Dip the filter into a 50 ml tube of Diff-Quick-1, 5×, 1 s each time.

(b) Dip the filter into a 50 ml tube of Diff-Quick-2, 5×, 1 s each time.

(c) Dip the filter into a 50 ml tube of Diff-Quick-3, 5×, 1 s each time.

(d) Dip the filter carefully into deionized filtered water several times to rinse off excess stain.

- Dry well by touching the edges of the filter to a paper towel.

(e) Remove the bottom clamp carefully, and lay the filter on a glass slide to partially dry.

(f) Lay the filter on a second glass slide. Gently pull out any wrinkles with the small forceps (from Neuro Probe accessory pack).

(g) The filter may be viewed and photographed as is or a drop of immersion oil may be added to cover the filter. Gently spread the oil on the filter with a smooth, blunt instrument to remove all bubbles and wrinkles.

4. View the slide using bright field microscopy (see Note 14).

5. Photograph three to five representative 20× fields of each well.

6. Count total cells in each field of view by hand or by using an image analysis software package such as ImageJ or Cell Profiler.

7. Graph results for analysis.

4 Notes

1. Cells used in this assay can be either freshly isolated, primary cultured cells or transformed cell lines. Cells of interest may include stem/progenitor cells, endothelial cells, and epithelial cells among others. The protocol described herein details an assay for the examination of PVSC chemotaxis.

2. Pore size is critical to the success of the assay and must be determined for each cell type.

3. The appropriate adhesion-promoting substrate (e.g., gelatin, fibronectin, laminin, collagen I, collagen IV) may be critical to the success of the assay and is cell type specific.

4. Density range varies for each cell type and is critical for the success of the assay. Most often, optimal density will be in the range of 10,000–250,000 cells per well.

5. Chemoattractants are substances with biologic activity, specifically, chemotactic potential. Common chemoattractant substances include pharmacologics, growth factors, secreted effector molecules, and degradation products of biologic scaffolds among others.

   Include a negative control consisting of migration media alone.

   When possible, include a positive control. The positive control may be a known specific chemoattractant (e.g., VEGF) at an appropriate concentration. For many cell types, basal media with 10–20 % serum is a good positive control.

6. The assay no longer requires sterility due to short incubation of migration periods (1–18 h).

7. Do not move filter once it is applied to prevent cross-contamination of wells and to minimize air bubble formation.

8. Maintain an even firm pressure on the top of the chamber while tightening the thumbnuts to prevent air bubbles from being drawn into the bottom wells. Thumbnuts should be finger-tight; do not use any tools to tighten.

9. The placement of cells into the chamber is performed on the bench top outside of the cell culture hood. Carefully add cells into each well. Eject the fluid with a rapid motion to dislodge air from the bottom of the well. Periodically resuspend the cells in the tube with a pipet to assure a single-cell suspension and even seeding densities.

10. Incubation time is critical to the success of the assay and must be determined for each cell type. Optimal incubation time for most cell types is between 2 and 4 h and is rarely longer than 6 h.

11. As the membrane filter is extremely thin, extreme care must be taken at each step from this point onward to assure that the filter remains intact and does not fold over on itself.

12. Care must be taken at each step from this point onward to assure that the filter remains intact and does not fold over on itself.

13. Alternatively, a fluorescent nuclear dye, such as Draq5 or Dapi, may be used to stain the filter for migrated cells.

14. Fluorescent microscopy must be used to image fluorescent nuclear dyes.

References

1. Reing JE, Zhang L, Myers-Irvin J, Cordero KE, Freytes DO, Heber-Katz E, Bedelbaeva K, McIntosh D, Dewilde A, Braunhut SJ, Badylak SF (2009) Degradation products of extracellular matrix affect cell migration and proliferation. Tissue Eng Part A 15(3): 605–614

2. Agrawal V, Siu BF, Chao H, Hirschi KK, Raborn E, Johnson SA, Tottey S, Hurley KB, Medberry CJ, Badylak SF (2012) Partial

Characterization of the Sox2+ Cell Population in an Adult Murine Model of Digit Amputation. Tissue Eng Part A 18(13):1454–1463

3. Crisan M, Yap S, Casteilla L, Chen CW, Corselli M, Park TS, Andriolo G, Sun B, Zheng B, Zhang L, Norotte C, Teng PN, Traas J, Schugar R, Deasy BM, Badylak S, Buhring HJ, Giacobino JP, Lazzari L, Huard J, Peault B (2008) A perivascular origin for mesenchymal stem cells in multiple human organs. Cell Stem Cell 3(3):301–313

4. Agrawal V, Johnson SA, Reing J, Zhang L, Tottey S, Wang G, Hirschi KK, Braunhut S, Gudas LJ, Badylak SF (2010) Epimorphic regeneration approach to tissue replacement in adult mammals. Proc Natl Acad Sci U S A 107(8):3351–3355

5. Tottey S, Corselli M, Jeffries EM, Londono R, Peault B, Badylak SF (2011) Extracellular matrix degradation products and low-oxygen conditions enhance the regenerative potential of perivascular stem cells. Tissue Eng Part A 17 (1–2):37–44

6. Tottey S, Johnson SA, Crapo PM, Reing JE, Zhang L, Jiang H, Medberry CJ, Reines B, Badylak SF (2011) The effect of source animal age upon extracellular matrix scaffold properties. Biomaterials 32(1):128–136

7. Badylak SF, Park K, Peppas N, McCabe G, Yoder M (2001) Marrow-derived cells populate scaffolds composed of xenogeneic extracellular matrix. Exp Hematol 29(11):1310–1318

8. Zantop T, Gilbert TW, Yoder MC, Badylak SF (2006) Extracellular matrix scaffolds are repopulated by bone marrow-derived cells in a mouse model of Achilles tendon reconstruction. J Orthop Res 24(6):1299–1309

9. Nieponice A, Gilbert TW, Johnson SA, Turner NJ, Badylak SF (2012) Bone marrow-derived cells participate in the long-term remodeling in a mouse model of esophageal reconstruction. J Surg Res 182(1):e1–e7

10. Brennan EP, Tang XH, Stewart-Akers AM, Gudas LJ, Badylak SF (2008) Chemoattractant activity of degradation products of fetal and adult skin extracellular matrix for keratinocyte progenitor cells. J Tissue Eng Regen Med 2 (8):491–498

11. Beattie AJ, Gilbert TW, Guyot JP, Yates AJ, Badylak SF (2009) Chemoattraction of progenitor cells by remodeling extracellular matrix scaffolds. Tissue Eng Part A 15(5):1119–1125

12. Valentin JE, Stewart-Akers AM, Gilbert TW, Badylak SF (2009) Macrophage participation in the degradation and remodeling of extracellular matrix scaffolds. Tissue Eng Part A 15 (7):1687–1694

13. Agrawal V, Kelly J, Tottey S, Daly KA, Johnson SA, Siu BF, Reing J, Badylak SF (2011) An isolated cryptic Peptide influences osteogenesis and bone remodeling in an adult Mammalian model of digit amputation. Tissue Eng Part A 17(23–24):3033–3044

14. Agrawal V, Tottey S, Johnson SA, Freund JM, Siu BF, Badylak SF (2011) Recruitment of progenitor cells by an extracellular matrix cryptic peptide in a mouse model of digit amputation. Tissue Eng Part A 17(19–20): 2435–2443

15. Gilbert TW, Stewart-Akers AM, Simmons-Byrd A, Badylak SF (2007) Degradation and remodeling of small intestinal submucosa in canine Achilles tendon repair. J Bone Joint Surg Am 89(3):621–630

16. Schober A, Weber C (2005) Mechanisms of monocyte recruitment in vascular repair after injury. Antioxid Redox Signal 7(9–10): 1249–1257

17. Yates CC, Bodnar R, Wells A (2011) Matrix control of scarring. Cell Mol Life Sci 68 (11):1871–1881

18. Janis JE, Kwon RK, Lalonde DH (2010) A practical guide to wound healing. Plast Reconstr Surg 125(6):230e–244e

19. Arnold L, Henry A, Poron F, Baba-Amer Y, van Rooijen N, Plonquet A, Gherardi RK, Chazaud B (2007) Inflammatory monocytes recruited after skeletal muscle injury switch into antiinflammatory macrophages to support myogenesis. J Exp Med 204(5):1057–1069

20. LaBarge MA, Blau HM (2002) Biological progression from adult bone marrow to mononucleate muscle stem cell to multinucleate muscle fiber in response to injury. Cell 111 (4):589–601

21. Palermo AT, Labarge MA, Doyonnas R, Pomerantz J, Blau HM (2005) Bone marrow contribution to skeletal muscle: a physiological response to stress. Dev Biol 279(2): 336–344

22. Lee JY, Qu-Petersen Z, Cao B, Kimura S, Jankowski R, Cummins J, Usas A, Gates C, Robbins P, Wernig A, Huard J (2000) Clonal isolation of muscle-derived cells capable of enhancing muscle regeneration and bone healing. J Cell Biol 150(5):1085–1100

23. Relaix F, Rocancourt D, Mansouri A, Buckingham M (2005) A Pax3/Pax7-dependent population of skeletal muscle progenitor cells. Nature 435(7044):948–953

24. Ten Broek RW, Grefte S, Von den Hoff JW (2010) Regulatory factors and cell populations involved in skeletal muscle regeneration. J Cell Physiol 224(1):7–16

Methods in Molecular Biology (2014) 1202: 111–119
DOI 10.1007/7651_2013_53
© Springer Science+Business Media New York 2013
Published online: 4 December 2013

Biomimetic Strategies Incorporating Enzymes into CaP Coatings Mimicking the In Vivo Environment

Ana M. Martins and Rui L. Reis

Abstract

Biomimetic calcium phosphate coating (CaP) methodology mimics the natural biomineralization process of bone, which involves two phases: nucleation and growth. In this chapter we present a successful method to coating biodegradable natural-origin materials with biomimetic CaP layer as a strategy to incorporate enzymes with the main aim of controlling and tailoring their degradation over time. This strategy seems to enhance the bone-bonding, osteoconductive, and osteoinductive properties of natural-origin biomaterials that per se do not present these characteristics.

Keywords: Natural-origin biomaterials, Biomimetics, CaP coatings, Enzymes

1   Introduction

Multidisciplinary approaches in tissue engineering have designed new materials inspired in the natural constituents of living organisms. The term "biomimetics" is used to describe a branch of science that seeks to produce such "bioinspired" materials for a variety of applications. This materials exhibit the same structural or functional properties of those observed in naturally occurring biological materials.

In the early 1990, Kokubo and colleagues (1, 2) proposed that the essential requirement for a biomaterial to bond to living bone is the formation of bone-like apatite on the surface of a biomaterial when implanted in vivo. This in vivo apatite formation can be reproduced in vitro using a simulated body fluid (SBF). SBF is a solution containing inorganic ion concentrations similar to those of human extracellular fluids without any cells and proteins (3). Under specific in vitro conditions, the de novo CaP layer consists of carbonate apatite known as "bone-like apatite" regarding its similarity to apatite present in bone.

In 1997, Reis et al. (4) adapted the methodology developed by Kokubo et al. (1) and used bioactive glass as a precursor of nucleation and growth of CaP films on starch-based polymers (4).

Biomimetic technique for coating biomaterials with bone-like apatite layer has been described in several publications by our group (4–13). This methodology mimics the natural biomineralization process, which involves two phases: nucleation and growth. The main advantage of the biomimetic technique is the use of physiological conditions, namely pH 7.4 and temperature of 37 °C, simulating the manner in which apatite is formed in bone. Also, this technique permits the incorporation of proteins without compromising their activity (5–7, 10, 11). CaP coatings were previously used as a carrier for enzymes as a strategy to control the degradation rate of biomaterials (5, 6, 10). Our group proposed a new concept of in situ pore formation (14). For that, we have used chitosan scaffolds that are biodegradable, biocompatible, and noncytotoxic (5–8, 14). In order to enhance the degradation rate of chitosan scaffolds and induce the formation of pores we have used lysozyme (5–8). This enzyme is present in the human body and it is responsible for the degradation of chitosan. Lysozyme was incorporated into CaP coatings onto the surface of chitosan scaffolds with the objective of developing a self-regulating biodegradable material with a capacity of in situ formation of pores (6). Most natural-origin scaffolds typically do not present adequate bone-bonding, osteoinductive, or osteoconductive properties for bone tissue engineering applications. One possible solution is to apply on the surface of scaffolds a CaP biomimetic coating with a composition similar to the major inorganic component of bone, hydroxyapatite. An improvement in the osteoconductivity of implants has been achieved by coating their surfaces with CaP layers (15–17).

Additional studies investigated the influence of a calcium phosphate (CaP) biomimetic coating on the osteogenic differentiation of rat bone marrow stromal cells (MSCs) and showed enhanced differentiation of rat MSCs seeded on the CaP-coated chitosan-based scaffolds with lysozyme incorporated to aid degradation (7). When compared with cells seeded onto control (uncoated) scaffolds, cells seeded onto CaP-coated chitosan-based scaffolds with incorporated lysozyme demonstrated at all culture times greater osteogenic differentiation, bone matrix production, and mineralization (7). Also, in vivo subcutaneous implantation was performed to evaluate the tissue response up to 12 weeks (5). At all time points no adverse tissue reaction was observed on scaffolds coated with CaP layer with or without lysozyme (5).

In this chapter, we propose a protocol to incorporate enzymes in CaP coatings in order to induce the osteogenic differentiation of MSCs, to enhance degradation of natural-origin biomaterials, and to promote osteoconductive potential of natural-based scaffolds.

Table 1
Reagents to prepare 2,000 mL of SBF 1 x (*see* Note 1)

Order	Reagents	Amount (g)
1	NaCl	16.072
2	NaHCO$_3$	0.704
3	KCl	0.450
4	KPO$_4$3H$_2$O	0.460
5	MgCl$_2$6H$_2$O	0.622
6	1.0 M-HCl	50 mL[a]
7	CaCl$_2$	0.586
8	Na$_2$SO$_4$	0.144
9	TRIS	12.236
10	1.0 M-HCl	1 mL

[a]The total amount of HCl that must be added is 75 mL. First it is added 50 mL and the remaining HCl solution must be added in the end, at the same time as TRIS

2 Materials

2.1 Preparation of SBF 1x solution

1. In order to prepare 2,000 mL of SBF, add 1,000 mL of distilled water in a beaker and a stirring bar. Set it in the water bath on the magnetic stirrer with heating (with controlled agitation and temperature) and cover it with plastic wrap.
2. Insert the electrode of the pH meter into the solution.
3. Heat the water in the beaker to 36.5 ± 1.5 °C under stirring.
4. Dissolve the aforementioned reagents following the order of Table 1 (*see* **Note 2**).
5. Add distilled water up to 1,900 mL in total (*see* **Note 3**).
6. Adjust the pH of the solution by dropping 1.0 M-HCl slowly to pH 7.4 and at 36.5 °C.
7. Transfer the solution to a volumetric flask of 2,000 mL. Wait until the temperature decreases to room temperature and adjust the exact volume adding distilled water (*see* **Note 4**).

2.2 Preparation of SBF 1.5x solution

Use the same procedure of Section 2.1. *See* Table 2.

2.3 Preparation of CaP Biomimetic Coatings on the Surface of Scaffolds

The method to prepare the biomimetic CaP coatings was based on the methodology previously developed by Abe and Kokubo et al. (1) and adapted by Reis et al. (4, 18) consisting in an impregnation of the materials with bioactive glass (Bioglass® 45S5, NovaMin

Table 2
Reagents to prepare 2,000 mL of SBF 1.5x (*see* Note 1)

Order	Reagents	Amount (g)
1	NaCl	24.108
2	NaHCO$_3$	1.056
3	KCl	0.675
4	KPO$_4$3H$_2$O	0.690
5	MgCl$_2$6H$_2$O	0.933
6	1.0 M-HCl	75 mL[a]
7	CaCl$_2$	0.879
8	Na$_2$SO$_4$	0.216
9	TRIS	18.354
10	1.0 M-HCl	1 mL

[a]The total amount of HCl that must be added is 112.5 mL. First it is added 75 mL and the remaining HCl solution must be added in the end, at the same time as TRIS

Table 3
Ion concentrations (mM) of blood plasma and SBF solution

	Na$^+$	K$^+$	Mg^{2+}	Ca^{2+}	Cl$^-$	HCO$_3^-$	HPO$_4^{3-}$	SO$_4^{2-}$
Blood plasma	142.0	5.0	1.5	2.5	103.0	27.0	1.0	0.5
SBF	142.0	5.0	1.5	2.5	147.8	4.2	1.0	0.5

Technology Inc or NOVABONE, Alachua, Florida, USA) followed by immersion in a simulated body fluid (SBF, 37 °C, pH 7.4), presenting ionic concentrations similar to human blood plasma (Table 3).

1. Sterilize scaffolds with ethylene oxide.
2. Sterilize Bioglass® 45S5 in ethanol solution (70 % v/v) and then let it dry inside the laminar flow cabinet (*see* **Note 5**).
3. Immerse the sterile scaffolds in a wet bed of Bioglass® using 50 ml tubes in a rotator for 6 h.
4. Add 10 mL of SBF 1× with lysozyme solution (*see* **Note 6**) to a 15 mL tubes.
5. Immerse the scaffolds in the aforementioned tubes at 37 °C for 7 days (nucleation stage) (*see* **Note 7**).
6. After 7 days prepare new 15 mL tubes with 10 mL of SBF 1.5×.

7. Wash the previous scaffolds with distilled water and immerse them in SBF 1.5× (prepared in the point 6) for 7 days at 37 °C in order to enhance CaP nuclei growth (growth stage).

8. After growth stage, wash scaffolds with distilled water and let them dry inside the laminar flow cabinet until surface characterization.

3 Methods

3.1 Scanning Electron Microscopy (SEM)

The morphology of the obtained CaP coatings should be analyzed by scanning electron microscopy. Prior to microstructure analysis, specimens should be coated with gold.

Figure 1 shows SEM images of the chitosan scaffold (Fig. 1a) and chitosan scaffolds coated with CaP after immersion in SBF 1× in the presence (Fig. 1c, e) and absence of lysozyme (Fig. 1b, d). At the nucleation stage it is possible to observe for both conditions (with and without lysozyme) the formation of apatite nuclei (Fig. 1b, c). Lysozyme incorporation in the nucleation stage did not affect significantly the morphology of the coatings (Fig. 1c). After 7 days (growth stage) for both conditions it is possible to observe that the size of CaP nuclei increased forming a dense and compact CaP film. This means that immersion in SBF 1.5× promotes the growth of CaP layer over time. At higher magnifications, after 7 days of growth, the morphology of the coatings is similar in the presence and absence of lysozyme, demonstrating the typical cauliflower morphology (Fig. 1d1, e1). Lysozyme incorporation into the coatings did not affect the morphology and growth of the coatings as a function of immersion time.

3.2 Fourier Transformed Infrared Spectroscopy (FTIR)

FTIR is performed to examine the chemical structure of the obtained CaP biomimetic coatings. Coatings should be scraped from the chitosan scaffolds, mixed with KBr and then formed into a disc in a press. All spectra should be obtained between 4,400 and 450 cm^{-1} at a 2 cm^{-1} resolution.

Figure 2A presents the infrared spectra of CaP coatings with and without lysozyme (control) at nucleation stage. Analysis of CaP coatings in absence of lysozyme shows the presence of phosphate (PO_4^{3-}) groups (bands at 1,055, 602, and 556 cm^{-1}). Moreover, the presence of carbonate (CO_3^{2-}) bands (1,492 and 1,441 cm^{-1}) was detected. When lysozyme was added in nucleation stage the same phosphate and carbonate groups were also detected (Fig. 2A). In the spectrum of CaP coating with lysozyme the appearance of amide I band (1,656 cm^{-1}) was observed indicating that lysozyme is incorporated in the coating. The amide I band represents the stretching vibrations of C=O bonds in the backbone of the proteins (10, 19). FTIR spectra of both CaP coatings, with and without

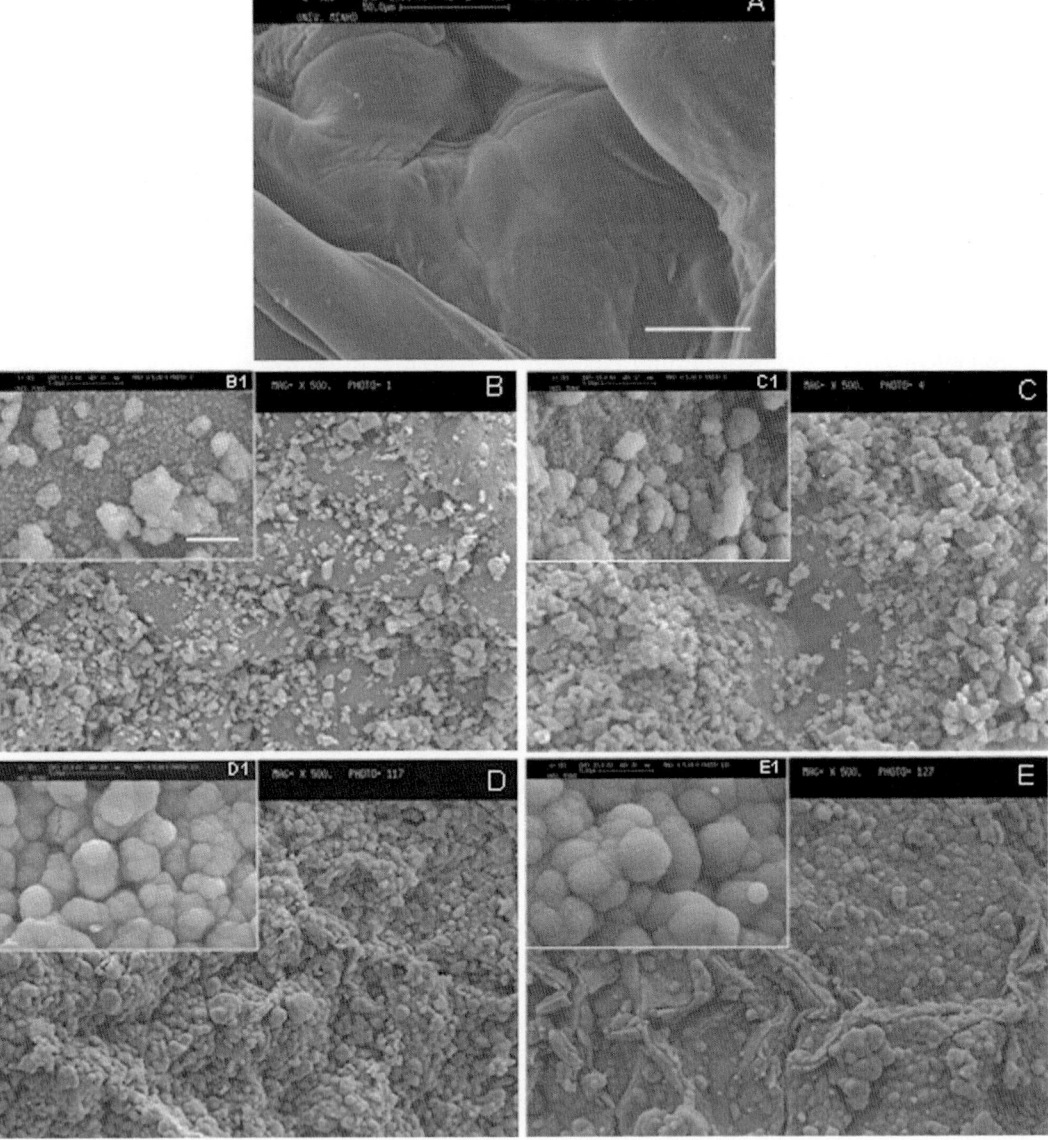

Fig. 1 SEM images of the surface of chitosan scaffolds: uncoated (**a**); coated with CaP layer at nucleation stage (**b**, **b1**) and after 7 days in the growth stage (**d**, **d1**) without enzymes; and coated with CaP layer with incorporated lysozyme at nucleation stage (**c**, **c1**) and after 7 days in the growth stage (**e**, **e1**). The scale bar for images **a–e** is 50 μm and for images **b1–e1** is 5 μm. From (6), with permission

lysozyme, confirmed an apatite layer formation on the surface of the scaffolds, with similar composition of the major mineral component of the bone (20).

3.3 Thin-Film X-Ray Diffraction (TF-XRD)

In addition, in order to characterize the crystalline/amorphous nature of the films, thin-film X-ray diffraction technique should be used. The data collection is performed by 2θ scan method

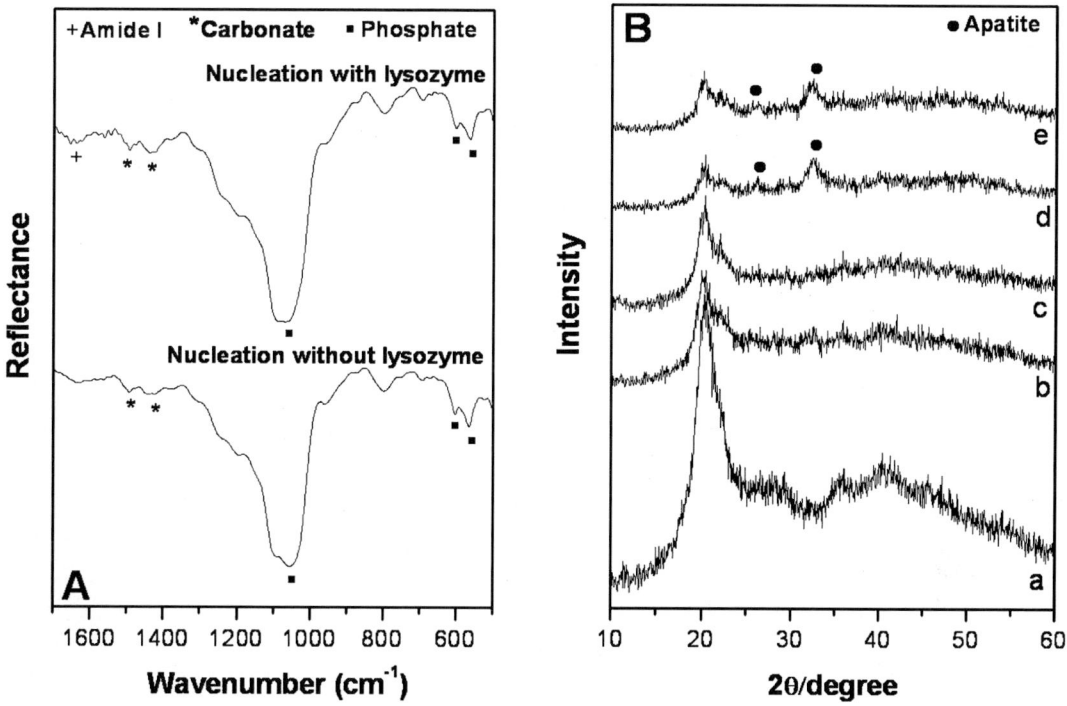

Fig. 2 FTIR spectra of CaP coatings, with and without incorporated lysozyme at nucleation stage (**A**) and TF-XRD spectra (**B**) of uncoated chitosan scaffolds (*a*); CaP-coated chitosan scaffolds at nucleation stage (*b*) and after 7 days of growth (*d*); CaP-coated chitosan with incorporated lysozyme at nucleation stage (*c*) and after 7 days of growth (*e*). From (6), with permission

with 1° as the incident beam angle using CuKα X-Ray line and a scan speed of 0.05°/min in 2θ.

TF-XRD analyses were performed to determine if the CaP coatings were crystalline or amorphous. Figure 2B shows TF-XRD patterns of CaP coatings in absence and presence of lysozyme. No crystalline peaks were detected for CaP-coated chitosan scaffolds with or without incorporated lysozyme at nucleation stage (Fig. 2Bb, Bc). After 7 days of growth, diffraction patterns of CaP coatings in the presence and absence of lysozyme, two characteristic peaks of hydroxyapatite are visible, which are confirmed by comparison to an XRD pattern of standard hydroxyapatite (JCPDS 9-432) (Fig. 2Bd, Be). These TF-XRD patterns confirm the formation of apatite layer due to the presence of apatite peaks. However, the apatite formed appears to be mainly amorphous. These results together with FTIR spectra suggest that the mineral formed is a carbonate apatite mineral similar to the major mineral component of bone.

4   Notes

1. Never dissolve several reagents simultaneously. Start to dissolve the next reagent when the previous one is completely dissolved.

2. Always make sure that the solution is colorless and transparent and without deposit on the surface of the beaker. If precipitation occurs, stop preparing SBF, discard the solution, and prepare a new one.

3. Add TRIS reagent slowly to raise the pH gradually up to 7.4. Make sure that the temperature of the solution is maintained at 36.5 ± 1.5 °C.

4. Store the SBF solution in the fridge. The SBF shall be used within 30 days after preparation.

5. All the subsequent works must be performed under sterile conditions in a laminar flow cabinet.

6. Prepare a solution of SBF 1× with lysozyme 1 g/L.

7. This stage allows for the formation of CaP nuclei while lysozyme is incorporated.

Acknowledgements

A.M.M. acknowledges "Fundação para a Ciência e Tecnologia" (FCT) for the Postdoctoral grant (SFRH/BPD/66897/2009) financed by POPH-QREN-Advanced Formation, and co-financed by Social European Fund and National Fund from MCTES.

References

1. Abe Y, Kokubo T, Yamamuro T (1990) Apatite coating on ceramics, metals and polymers utilizing a biological process. J Mater Sci Mater Med 1:233–238

2. Kokubo T (1991) Bioactive glass-ceramics - properties and applications. Biomaterials 12:155–163

3. Ohtsuki C, Kamitakahara M, Miyazaki T (2007) Coating bone-like apatite onto organic substrates using solutions mimicking body fluid. J Tissue Eng Regen Med 1:33–38

4. Reis RL et al (1997) Treatments to induce the nucleation and growth of apatite-like layers on polymeric surfaces and foams. J Mater Sci Mater Med 8:897–905

5. Martins AM et al (2012) Gradual pore formation in natural origin scaffolds throughout subcutaneous implantation. J Biomed Mater Res A 100A:599–612

6. Martins AM et al (2009) Chitosan scaffolds incorporating lysozyme into CaP coatings produced by a biomimetic route: A novel concept for tissue engineering combining a self-regulated degradation system with in situ pore formation. Acta Biomater 5:3328–3336

7. Martins AM et al (2009) Natural stimulus responsive scaffolds/cells for bone tissue engineering: influence of lysozyme upon scaffold degradation and osteogenic differentiation of cultured marrow stromal cells induced by CaP coatings. Tissue Eng Part A 15:1953–1963

8. Martins AM et al (2006) Lysozyme incorporation in biomimetic coated chitosan scaffolds: development and behaviour in contact with osteoblastic-like cells. Tissue Eng 12:1018–1019

9. Araujo JV et al (2007) Surface controlled biomimetic coating of polycaprolactone nanofiber mesh architectures. Tissue Eng 13:1759–1759

10. Azevedo HS et al (2005) Incorporation of proteins and enzymes at different stages of the preparation of calcium phosphate coatings on a degradable substrate by a biomimetic methodology. Mater Sci Eng C Biomim Supramol Syst 25:169–179

11. Azevedo HS, Leonor IB, Reis RL (2006) Incorporation of proteins with different isoelectric points into biomimetic Ca-P coatings: a new approach to produce hybrid coatings with tailored properties. Bioceramics 18(Pts 1 and 2):309–311, 755–758

12. Leonor IB et al (2004) Learning from nature how to design biomimetic calcium-phosphate coatings. Learn Nat How Design New Implantable Biomater Biominer Fundam Biomim Mater Process Routes 171:123–150

13. Leonor IB, Reis RL (2003) An innovative auto-catalytic deposition route to produce calcium-phosphate coatings on polymeric biomaterials. J Mater Sci Mater Med 14:435–441

14. Martins AM et al (2008) Natural origin scaffolds with in situ pore forming capability for bone tissue engineering applications. Acta Biomater 4:1637–1645

15. Habibovic P et al (2005) Biological performance of uncoated and octacalcium phosphate-coated Ti6A14V. Biomaterials 26:23–36

16. Liu Y, de Groot K, Hunziker EB (2005) BMP-2 liberated from biomimetic implant coatings induces and sustains direct ossification in an ectopic rat model. Bone 36:745–757

17. Habibovic P, de Groot K (2007) Osteoinductive biomaterials—properties and relevance in bone repair. J Tissue Eng Regen Med 1:25–32

18. Oliveira AL et al (1999) Surface modification tailors the characteristics of biomimetic coatings nucleated on starch-based polymers. J Mater Sci Mater Med 10:827–835

19. Xie J et al (2002) FTIR/ATR study of protein adsorption and brushite transformation to hydroxyapatite. Biomaterials 23:3609–3616

20. Mann S (2001) Biomineralization. Principles and concepts in bioinorganic materials chemistry. Oxford University Press, Oxford

Methods in Molecular Biology (2014) 1202: 121–130
DOI 10.1007/7651_2014_84
© Springer Science+Business Media New York 2014
Published online: 30 May 2014

Fabrication of Biofunctionalized, Cell-Laden Macroporous 3D PEG Hydrogels as Bone Marrow Analogs for the Cultivation of Human Hematopoietic Stem and Progenitor Cells

Lisa Rödling, Annamarija Raic, and Cornelia Lee-Thedieck

Abstract

In vitro proliferation of hematopoietic stem cells (HSCs) is yet an unresolved challenge. Found in the bone marrow, HSCs can undergo self-renewing cell division and thereby multiply. Recapitulation of the bone marrow environment in order to provide the required signals for their expansion is a promising approach.

Here, we describe a technique to produce biofunctionalized, macroporous poly(ethylene glycol) diacrylate (PEGDA) hydrogels that mimic the spongy 3D architecture of trabecular bones, which host the red, blood-forming bone marrow. After seeding these scaffolds with cells, they can be used as simplified bone marrow analogs for the cultivation of HSCs. This method can easily be conducted with standard laboratory chemicals and equipment. The 3D hydrogels are produced via salt leaching and biofunctionalization of the material is achieved by co-polymerizing the PEGDA with an RGD peptide. Finally, cell seeding and retrieval are described.

Keywords: Hematopoietic stem and progenitor cells, Biomimetic material, 3D scaffolds, Bone marrow analog, Stem cell niche

1 Introduction

Hematopoietic stem cells (HSCs) are used in the clinics to treat patients with hematological malignancies. Typically, patients receive autologous or allogeneic HSC transplants extracted from bone marrow or from peripheral blood after pharmacological stem cell mobilization. This kind of HSC supply, however, is limited. In addition, umbilical cord blood has proven to be an effective and noninvasive source for HSCs (1). The number of HSCs in one cord blood unit is strongly limited and only sufficient to transplant children or adults with very low body weight (2). Therefore, development of an efficient strategy to propagate healthy HSCs ex vivo would be an elegant way to increase the potential of umbilical cord blood as a HSC source for clinical significance.

When HSCs are cultured in conventional cell culture conditions, they rapidly lose their stem cell potential, i.e., the ability to self-renew and to differentiate into all different blood cell lineages (3).

This is in sharp contrast to their behavior in their natural microenvironment—their niche. Here HSCs divide and maintain their stem cell properties over an individuals' life time (4). Stem cell niches are highly specialized with respect to stem cell maintenance and differentiation. In their niche, HSCs receive the required stimuli to pursue their function. The stimuli are provided by (I) direct and indirect contact and communication with other niche cells such as osteoblasts, endothelial cells, or mesenchymal stem/stromal cells (MSCs) (5–10), (II) by interaction with the extracellular matrix (ECM) (11, 12), (III) by soluble/diffusible factors such as growth factors or hormones, and (13) (IV) by (bio)physical parameters including matrix stiffness, or the three-dimensional architecture (14–18). Therefore, to increase the success of in vitro hematopoietic stem and progenitor cell (HSPC) propagation, the bone marrow conditions need to be mimicked. Culturing on 2D state-of-the-art tissue culture plastic or in suspension is not sufficient. Accordingly, three-dimensional (3D) culture strategies clearly favor HSPC proliferation compared to 2D cultures (19). Current 3D HSPC culture systems can be roughly subdivided into four different classes: (I) polymeric hydrogels into which HSPCs can be incorporated (14, 20, 21), (II) nano-fibrous networks (22), (III) quasi-3D micro-wells (23, 24), and (IV) macroporous biomaterials (25, 18, 26) that recapitulate the architectural features of trabecular bones that host the red bone marrow, including HSC niches.

Here we describe a straightforward, easy-to-use strategy for the production of a macroporous poly(ethylene glycol) (PEG)-based bone marrow analog that provides the opportunity to mimic four crucial aspects of the HSC niche: the 3D architecture, cell-adhesive signals of the ECM, interaction with supporting cells such as MSCs, and stimulation by soluble factors such as cytokines (16) (Fig. 1a). We give detailed instruction on the production of the macroporous hydrogels via salt leaching, biofunctionalization with RGD-peptides to allow cell adhesion, seeding of the material with HSPCs and MSCs, culture of the cells in the 3D scaffolds, and finally on how to retrieve the cultured HSPCs.

2   Materials

Prepare all solutions using Milli-Q water (prepared by purifying deionized water to attain a sensitivity of 18 MΩ cm at 25 °C) and molecular biology grade reagents. Store all reagents and solutions at room temperature unless otherwise specified.

2.1   Components for Fabrication of Macroporous PEGDA Hydrogels

1. Poly(ethylene glycol) diacrylate (PEGDA) (see **Note 1**): $M_r = 6,000$. Store at $-20\ °C$, protect from light.

2. RGD (see **Note 2**): RGDSK-PEG-acrylate; RGDSK peptide that carries a PEG spacer functionalized with an acrylate moiety

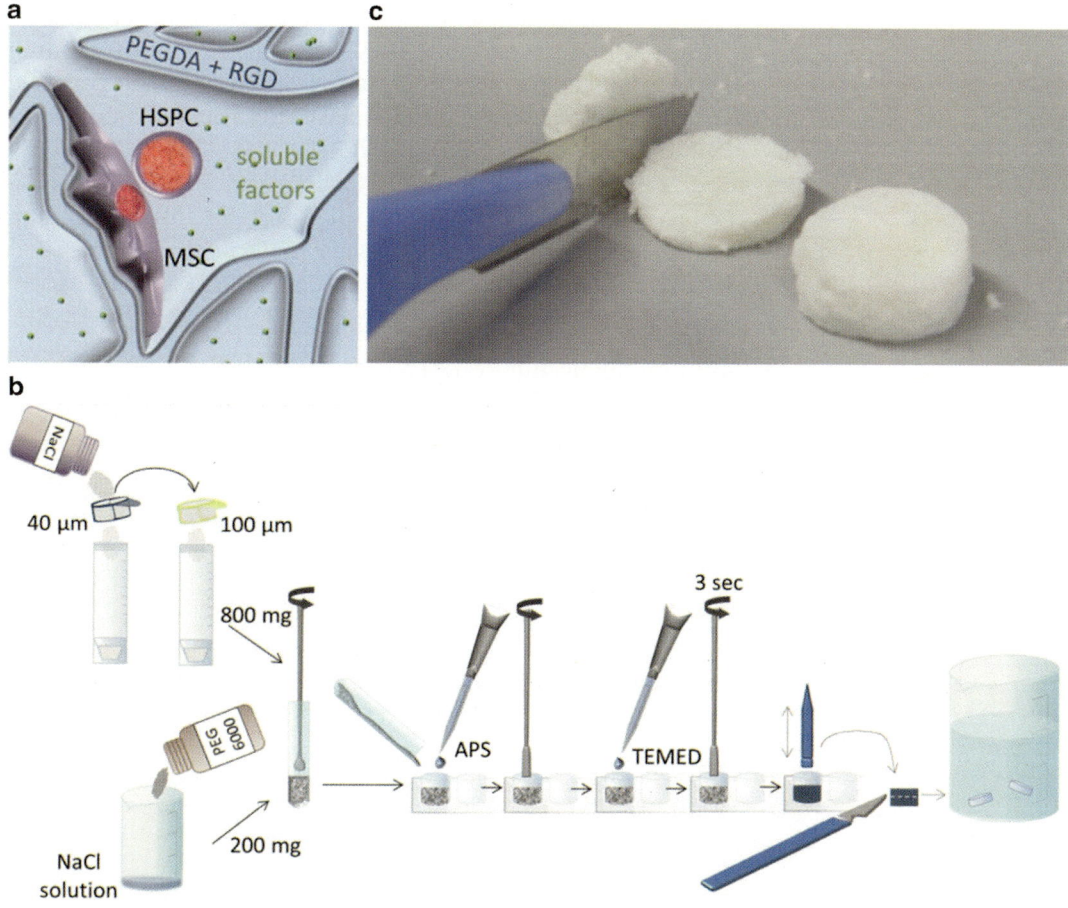

Fig. 1 Overview of macroporous hydrogels and their production. (**a**) Schematic representation of a macroporous PEGDA hydrogel containing co-polymerized RGD peptide, seeded with HSPCs together with MSCs, and filled with culture medium containing soluble factors. (**b**) Schematic illustration of the fabrication process of macroporous PEGDA hydrogels. (**c**) Freshly prepared PEGDA hydrogels (*right*) are cut into two discs using a scalpel (*left*)

at the ε-amino group of the lysine residue. Store at 4 °C as 500 μM stock in water.

3. Saturated aqueous NaCl solution: Warm up water to 50 °C and add NaCl (Merck, Darmstadt, Germany) while stirring until the NaCl crystals do not dissolve (solution is saturated). Let the solution cool down to room temperature. Undissolved NaCl crystals will sediment on the bottom. Only use the clear saturated NaCl solution.

4. Nylon cell strainers (BD Falcon, Heidelberg, Germany): 40 and 100 μm mesh sizes.

5. Parafilm.

6. Size-selected NaCl crystals (40–100 μm): place a 40 and a 100 μm cell strainer on 50 mL centrifuge tubes. Pestle NaCl

crystals using a mortar, transfer the crystals onto the 40 µm Nylon cell strainer, and sieve by shaking the tube at 600 rpm on a Thermomixer for 20 min. Transfer the crystals bigger than 40 µm from the top of the sieve onto the 100 µm Nylon cell strainer and sieve those collected crystals by shaking the tube at 600 rpm for 20 min. Collect the 40–100 µm NaCl crystals inside the centrifuge tube. Repeat until there are enough 40–100 µm NaCl crystals for one gel (*see* **Note 3**) (Fig. 1b top left).

7. 10 mL glass beaker or glass sample vial.

8. Ammonium persulfate (APS): 10 % (w/v) in water. Prepare 70 µL aliquots and store at −20 °C for up to 1 month.

9. *N*,*N*,*N'*,*N'*-Tetramethylethane-1,2-diamine (TEMED). Store at 4 °C.

10. Round stamp: e.g., a blue tip for 1,000 µL pipets that has parafilm span over its large opening.

11. Magnetic stirrer.

12. Small magnetic stir bars (8 mm).

13. 5 mL round bottom test tube (e.g., flow cytometry tube).

14. 48-well plate.

15. Scalpel.

16. 1,000 mL glass beakers.

17. Orbital shaker.

2.2 Components for Seeding/Culturing of Cells

Where possible use sterile components, e.g., cell culture equipment.

1. Flat plastic tweezers (e.g., K35a plastic tweezers; Rubis, Stabio, Switzerland).

2. Ethanol (*see* **Note 4**).

3. 10 cm glass petri dish.

4. Freeze dryer (e.g., Imago 500; Jumo, Fulda, Germany).

5. 24-well plate.

6. CD34[+] hematopoietic stem and progenitor cells (HSPCs) isolated from human umbilical cord blood by immunomagnetic cell separation (MACS; Miltenyi Biotec, Bergisch Gladbach, Germany).

7. Mesenchymal stem/stromal cells (MSCs) from bone marrow.

8. Commercially available HSC expansion medium, e.g., HematoStem SF Kit (PAA, Pasching, Austria).

9. Penicillin/streptomycin: 100× (Gibco, Darmstadt, Germany).

10. L-Glutamine: 200 mM (Gibco).

2.3 Components for Cell Retrieval

Where possible use sterile components, e.g., cell culture equipment.

1. PBS without Ca^{2+} and Mg^{2+}.
2. 35 mm petri dish.
3. Trypsin/EDTA: 0.05 % (Gibco).
4. Fetal Bovine Serum (Sigma-Aldrich, Taufkirchen, Germany).
5. Thermomixer.

3 Methods

Handle the gels with flat plastic tweezers and carry out all procedures at room temperature unless otherwise specified.

3.1 Fabrication of Macroporous 3D PEGDA Hydrogels

1. Dissolve 333 mg PEGDA in 1 mL of saturated aqueous NaCl solution in a 10 mL glass beaker and stir at 130 rpm for 1 h in the dark. Add 40 μL 500 μM RGD (final concentration ~ 20 μM) and stir until further use, minimum for 15 min (*see* **Note 5**).

2. Weigh 800 mg of the 40–100 μm NaCl crystals into a 5 mL round bottom test tube and add 200 mg of the PEGDA/RGD solution (*see* **Note 6**) (Fig. 1b). Mix the NaCl crystals and PEGDA/RGD solution well using a spatula. Transfer the mass into a well of a 48-well plate.

3. For polymerization of the PEGDA, add 60 μL of 10 % APS (w/v) and mix well using a small spatula (*see* **Note 7**). Add 10 μL of TEMED, then immediately mix well and press down the mass using the round stamp in order to form a continuous, flat gel (*see* **Notes 8** and **9**).

4. After 20 min, carefully detach and remove the gel from the well using a small spatula. Cut the gel in half so that two disc shaped gels of 2–3 mm width are obtained (*see* **Note 10**) (Fig. 1b, c).

5. Place the gels in 800 mL of water in a 1,000 mL glass beaker and shake on an orbital shaker to leach out the incorporated NaCl crystals. Exchange the water every hour on the first day and the following 2 days three times a day (*see* **Note 11**). After 3 days, a completely swollen, macroporous hydrogel biofunctionalized with RGD peptide is obtained.

3.2 Seeding Macroporous Hydrogels with Cells

The preparation of the macroporous hydrogels for cell seeding is schematically illustrated in Fig. 2a. Figure 2b gives an impression on the appearance of the gels at each production step.

1. For dehydration of the hydrogel, place it in ethanol with increasing concentration (50, 60, 70, 80, 90, and 95 % (v/v) ethanol in water and $2\times$ 100 % (v/v) ethanol). Incubate the gel for 10 min in each concentration.

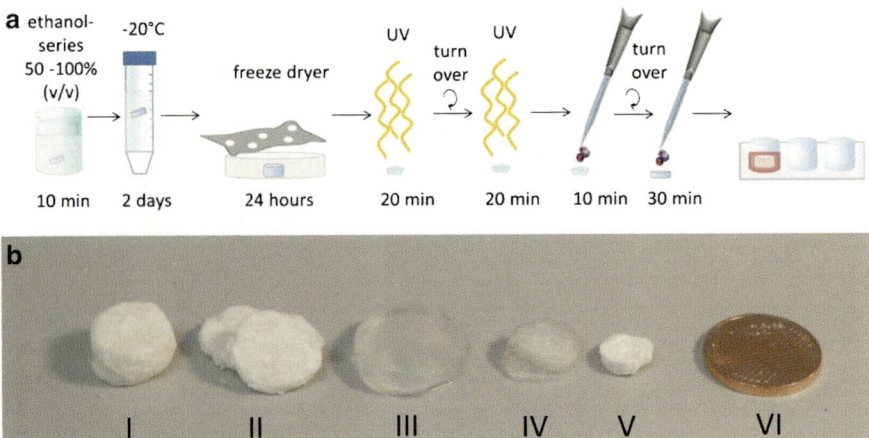

Fig. 2 Seeding of the macroporous hydrogels and optical appearance. (**a**) Schematic illustration of the dehydration, freeze drying, sterilization, and cell seeding process of macroporous PEGDA hydrogels. (**b**) The different appearances of PEGDA hydrogels occurring during individual experimental steps: (*I*) freshly polymerized, (*II*) freshly polymerized and cut in half, (*III*) swollen, (*IV*) for 2 days frozen in 100 % (v/v) ethanol, (*V*) freeze dried hydrogel, and (*VI*) 1 cent euro coin for size comparison

2. Put the gel into a 50 mL centrifuge tube with 100 % (v/v) ethanol and freeze the gel for 2 days at −20 °C (*see* **Note 12**).

3. Transfer the gel to a 10 cm glass petri dish, cover with parafilm, and punch small holes in the parafilm to allow air exchange. Put it into a freeze dryer and freeze-dry the gel for 24 h (*see* **Note 13**).

4. Put the freeze-dried gel into a desiccator for 4 h and let it come to room temperature.

5. Sterilize the dried gel by UV irradiation under the sterile bench for 20 min. Then turn over the gel and irradiate for another 20 min with UV light. Sterile techniques should now be utilized.

6. Prepare cell suspensions of 1.25×10^5 HSPCs together with 5×10^5 MSCs in 100 μL HSC expansion medium (e.g., HematoStem SF Kit) supplemented with 1 % (v/v) penicillin/streptomycin and 2 mM L-Glutamine.

7. Transfer the gel into a well of a 24-well plate.

8. Pipet 50 μL of the cell suspension onto the dry gel and wait for 10 min to allow the suspension to be soaked up by the gel.

9. Turn over the gel and apply the remaining 50 μL of the cell suspension onto the gel. Wait for 30 min.

10. Place the gel into a fresh well and fill up the well with 2 mL of HSC expansion medium containing 1 % (v/v) penicillin/streptomycin and 2 mM L-Glutamine and culture the cells in a humidified incubator at 37 °C and 5 % CO_2.

11. Exchange the medium every second day (*see* **Note 14**).

3.3 Retrieving Cells Cultured in the Macroporous Hydrogel

A schematic overview of the consecutive steps of cell retrieval is illustrated in Fig. 3.

1. After 4 days in culture, remove the culture medium and rinse the gel gently with PBS (*see* **Note 15**).

2. Place the gel into a 50 mL centrifuge tube, add 2 mL PBS, and wash by shaking at 700 rpm and 37 °C for 15 min on a Thermomixer. Transfer the gel into a new centrifuge tube and collect the used tube with the PBS containing cells.

3. Add 2 mL of fresh PBS to the gel, wash by shaking at 700 rpm and 37 °C for 15 min. Centrifuge the tube for 5 min at 600 × *g*, place the gel into a 35 mm petri dish, and collect the tube with PBS containing cells.

4. Cut the gel in the 35 mm petri dish into small pieces using a scalpel.

5. Transfer the gel pieces into a 50 mL centrifuge tube, add 2 mL of Trypsin/EDTA, and shake at 700 rpm and 37 °C for 30 min on a Thermomixer. Centrifuge the tube for 5 min at 600 × *g*.

6. Place the gel pieces into a new tube with 2 mL PBS and shake it at 700 rpm for 15 min. Transfer the gel pieces into a new tube with 2 mL of PBS, collect the old tube with PBS containing cells, and repeat this step.

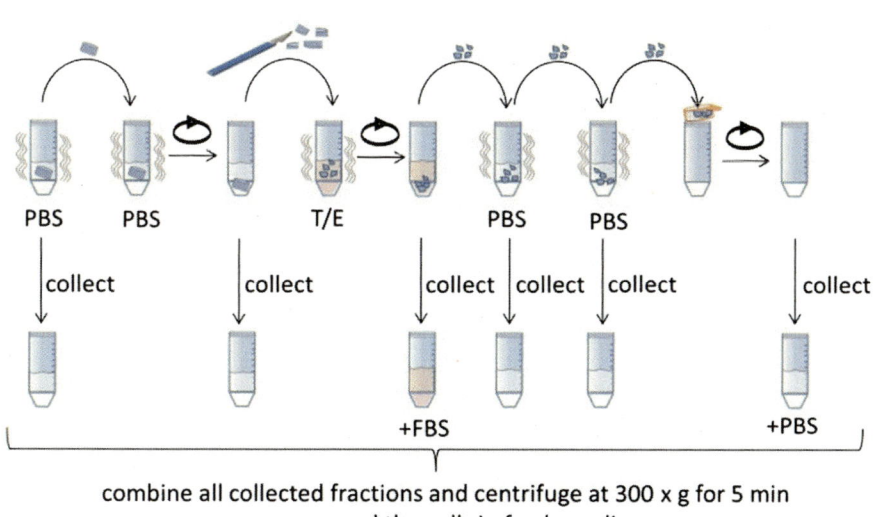

Fig. 3 Schematic illustration of the successive retrieval of cultured HSPCs from the macroporous PEGDA hydrogels. The cells are harvested from the hydrogels by successive washing, trypsinization, and centrifugation steps as illustrated above and described in Section 3.3. *T/E* Trypsin/EDTA

7. While the gel pieces are washed twice in PBS (step 6), add 1 mL FBS to the tube with the Trypsin/EDTA fraction from step 5 and mix to inhibit Trypsin.

8. Put the gel pieces on a Nylon cell strainer with 100 μm mesh size. Place the cell strainer on a 50 mL centrifuge tube and centrifuge the tube for 5 min at 300 × g.

9. Remove the cell strainer, add 2 mL PBS into the centrifuge tube, resuspend pelleted cells, and collect the PBS.

10. Combine all collected fractions (containing the cells) in one tube and centrifuge at 300 × g for 5 min. Resuspend the pelleted cells in 100 μL of fresh media.

4   Notes

1. PEGDA was synthesized as described previously (27).

2. RGDSK-PEG-acrylate was synthesized as described previously (16).

3. Clean the cell strainers from time to time to prevent blocking of pores using a soft brush or interdental tooth brush. Cover the cell strainer with parafilm in order to prevent crystals being lost. Check every 5 min that the crystals are in contact with the mesh and do not accumulate at the walls of the cell strainer.

4. Use high purity ethanol and not technical ethanol to avoid unwanted components in the gels.

5. Less RGD-peptide will result in reduced cell adhesion whereas more RGD-peptide does not enhance cell adhesion (16).

6. Dissolved NaCl crystals will precipitate out of solution when PEGDA is added. These crystals were reported to be smaller than 20 μm in size (28) and are meant to be present in the solution because they promote the interconnectivity of the pores. Keep them distributed evenly in the gel precursor by stirring the solution right up to the time when it is needed for the hydrogel production.

7. APS also serves to provide moisture, which allows the PEGDA/RGD solution and NaCl crystals to be more easily mixed.

8. Working quickly is essential, since the gel polymerizes very rapidly due to the relatively high amounts of APS and TEMED used. You have about 3 s to mix the TEMED and to press the gel precursor into shape before the polymerization stops.

9. Note that if you increase or decrease the RGD concentration for any reason you have to adjust the APS and TEMED concentration.

10. Parts of the gel can be friable, in which case, the TEMED was not distributed evenly throughout the precursor mass resulting in incomplete polymerization.

11. When the NaCl crystals leach out from the gel, the gel will start to rise in the water and sinks back down when the crystals are completely removed. The gel is very soft and unstable until the salt is completely leached out. Do not remove the gel earlier.

12. The hydrogel shrinks to about a third when kept in 100 % ethanol (v/v) (Fig. 2b).

13. The gel shrinks further when freeze dried and turns white when it is completely dry.

14. Due to the cultivation conditions the surface of the seeded hydrogel changes to an uneven surface structure.

15. With PBS cells on the surface of the hydrogel are washed away.

Acknowledgements

This work was supported by the BioInterfaces programme of the Helmholtz-Association, the "Brigitte Schlieben-Lange-Programm", the "Käthe und Josef Klinz-Stiftung", and the BMBF NanoMatFutur programme (FKZ: 13N12968).

References

1. Rao M, Ahrlund-Richter L, Kaufman DS (2012) Concise review: cord blood banking, transplantation and induced pluripotent stem cell: success and opportunities. Stem Cells 30 (1):55–60

2. Monti M, Perotti C, Del Fante C et al (2012) Stem cells: sources and therapies. Biol Res 45 (3):207–214

3. Walasek MA, van Os R, de Haan G (2012) Hematopoietic stem cell expansion: challenges and opportunities. Ann N Y Acad Sci 1266:138–150

4. Morrison SJ, Spradling AC (2008) Stem cells and niches: mechanisms that promote stem cell maintenance throughout life. Cell 132 (4):598–611

5. Calvi LM, Adams GB, Weibrecht KW et al (2003) Osteoblastic cells regulate the haematopoietic stem cell niche. Nature 425 (6960):841–846

6. Ding L, Morrison SJ (2013) Haematopoietic stem cells and early lymphoid progenitors occupy distinct bone marrow niches. Nature 495(7440):231–235

7. Kiel MJ, Radice GL, Morrison SJ (2007) Lack of evidence that hematopoietic stem cells depend on N-cadherin-mediated adhesion to osteoblasts for their maintenance. Cell Stem Cell 1(2):204–217

8. Kiel MJ, Yilmaz OH, Iwashita T et al (2005) SLAM family receptors distinguish hematopoietic stem and progenitor cells and reveal endothelial niches for stem cells. Cell 121 (7):1109–1121

9. Mendez-Ferrer S, Michurina TV, Ferraro F et al (2010) Mesenchymal and haematopoietic stem cells form a unique bone marrow niche. Nature 466(7308):829–834

10. Zhang J, Niu C, Ye L et al (2003) Identification of the haematopoietic stem cell niche and control of the niche size. Nature 425 (6960):836–841

11. Klein G (1995) The extracellular matrix of the hematopoietic microenvironment. Experientia 51(9–10):914–926

12. Prosper F, Verfaillie CM (2001) Regulation of hematopoiesis through adhesion receptors. J Leukoc Biol 69(3):307–316

13. Wang LD, Wagers AJ (2011) Dynamic niches in the origination and differentiation of haematopoietic stem cells. Nat Rev Mol Cell Biol 12 (10):643–655

14. Choi JS, Harley BA (2012) The combined influence of substrate elasticity and ligand density on the viability and biophysical properties of hematopoietic stem and progenitor cells. Biomaterials 33(18):4460–4468

15. Holst J, Watson S, Lord MS et al (2010) Substrate elasticity provides mechanical signals for the expansion of hemopoietic stem and progenitor cells. Nat Biotechnol 28 (10):1123–1128

16. Raic A, Rödling L, Kalbacher H et al (2014) Biomimetic macroporous PEG hydrogels as 3D scaffolds for the multiplication of human hematopoietic stem and progenitor cells. Biomaterials 35(3):929–940

17. Lee-Thedieck C, Rauch N, Fiammengo R et al (2012) Impact of substrate elasticity on human hematopoietic stem and progenitor cell adhesion and motility. J Cell Sci 125 (16):3765–3775

18. Lee J, Kotov NA (2009) Notch ligand presenting acellular 3D microenvironments for ex vivo human hematopoietic stem-cell culture made by layer-by-layer assembly. Small 5 (9):1008–1013

19. Lee J, Cuddihy MJ, Kotov NA (2008) Three-dimensional cell culture matrices: state of the art. Tissue Eng Part B Rev 14(1):61–86

20. Leisten I, Kramann R, Ventura Ferreira MS et al (2012) 3D co-culture of hematopoietic stem and progenitor cells and mesenchymal stem cells in collagen scaffolds as a model of the hematopoietic niche. Biomaterials 33 (6):1736–1747

21. Ventura Ferreira MS, Jahnen-Dechent W, Labude N et al (2012) Cord blood-hematopoietic stem cell expansion in 3D fibrin scaffolds with stromal support. Biomaterials 33 (29):6987–6997

22. Feng Q, Chai C, Jiang XS et al (2006) Expansion of engrafting human hematopoietic stem/progenitor cells in three-dimensional scaffolds with surface-immobilized fibronectin. J Biomed Mater Res A 78(4):781–791

23. Kurth I, Franke K, Pompe T et al (2011) Extracellular matrix functionalized microcavities to control hematopoietic stem and progenitor cell fate. Macromol Biosci 11 (6):739–747

24. Lutolf MP, Doyonnas R, Havenstrite K et al (2009) Perturbation of single hematopoietic stem cell fates in artificial niches. Integr Biol (Camb) 1(1):59–69

25. Di Maggio N, Piccinini E, Jaworski M et al (2011) Toward modeling the bone marrow niche using scaffold-based 3D culture systems. Biomaterials 32(2):321–329

26. Mortera-Blanco T, Mantalaris A, Bismarck A et al (2011) Long-term cytokine-free expansion of cord blood mononuclear cells in three-dimensional scaffolds. Biomaterials 32 (35):9263–9270

27. Aydin D, Louban I, Perschmann N et al (2010) Polymeric substrates with tunable elasticity and nanoscopically controlled biomolecule presentation. Langmuir 26 (19):15472–15480

28. Chiu YC, Larson JC, Isom A Jr et al (2010) Generation of porous poly(ethylene glycol) hydrogels by salt leaching. Tissue Eng Part C Methods 16(5):905–912

Methods in Molecular Biology (2014) 1202: 131–148
DOI 10.1007/7651_2013_35
© Springer Science+Business Media New York 2013
Published online: 12 February 2014

Extracellular Matrix Mimetic Peptide Scaffolds for Neural Stem Cell Culture and Differentiation

Busra Mammadov, Mustafa O. Guler, and Ayse B. Tekinay

Abstract

Self-assembled peptide nanofibers form three-dimensional networks that are quite similar to fibrous extracellular matrix (ECM) in their physical structure. By incorporating short peptide sequences derived from ECM proteins, these nanofibers provide bioactive platforms for cell culture studies. This protocol provides information about preparation and characterization of self-assembled peptide nanofiber scaffolds, culturing of neural stem cells (NSCs) on these scaffolds, and analysis of cell behavior. As cell behavior analyses, viability and proliferation of NSCs as well as investigation of differentiation by immunocytochemistry, qRT-PCR, western blot, and morphological analysis on ECM mimetic peptide nanofiber scaffolds are described.

Keywords: Peptide nanofibers, Scaffolds, Self-assembly, Hydrogels, Neural stem cells

1 Introduction

Neural stem cells (NSCs) are the stem cells of the central nervous system (CNS), which are present in both embryonic and adult brain. In contrary to other neural cells of CNS, NSCs have proliferative ability. They demonstrate not only self-renewing ability but also potential to differentiate into neurons, astrocytes, and oligodendrocytes (1). However, NSCs are very low in number and are localized to very restricted areas in brain (such as dentate gyrus and subventricular zone in adult human brain), which make them insufficient to regenerate the brain after any neural degeneration. Transplantation of either NSCs or differentiated cells from NSCs is considered to be a promising approach for CNS regeneration (2). Obtaining NSCs for such a therapy has been the pitfall of this approach as they can only be obtained from adult cadavers or aborted fetuses, both of which can be considered as allotransplants bearing risk of immune rejection. In addition, use of fetal cells raises ethical concerns about such a therapy. Fortunately, iPSC-derived

NSCs have recently been introduced, solving both the donor shortage problem and immune rejection risk, as they are the own cells of the patient (3). They have been proved to have similar self-renewal and differentiation capacity to the NSCs isolated from CNS (4). Besides, generation of NSCs directly from fibroblasts is also found to be possible by skipping the iPSC step. These cells are even safer for clinical use as they are proved to have no teratogenic risk upon transplantation (5, 6).

Although it is promising to transplant NSCs to the damaged area of the brain, it is not as successful as desired when used alone. This is mainly due to the non-permissive environment in the degenerated area which inhibits regeneration (7). Hence, transplantation within a permissive scaffold, which supports adhesion and proliferation of NSCs or axonal extension by neurons differentiated from NSCs, would be a better alternative to transplanting NSCs alone.

Self-assembled peptide nanofibers are powerful tools for producing bioactive scaffolds that are tailored according to the demands of the specific cells. Peptide amphiphiles (PA), which are molecules consisting of a hydrophobic alkyl chain conjugated to a peptide sequence, can be used as building blocks of these scaffolds. Peptide part of the PA molecules usually contains an amino acid sequence containing β-sheet-prone amino acids such as valine or alanine along with some charged amino acids and an epitope sequence for bioactivity (8). Charged amino acids are important both for solubility of the peptide and for nanofiber formation. PA molecules of opposite charges come together by electrostatic interactions in aqueous environment along with the hydrophobic interactions of alkyl groups. Through these interactions, peptide nanofibers, that are 8–10 nm in diameter, are produced. These nanofibers are actually cylindrical micelles where hydrophobic tails are embedded in the nanofibers and hydrophilic peptide sequences containing the epitope remain on the periphery of the nanofibers. During formation of the nanofibers, water is trapped in the peptide nanofiber networks leading to formation of a hydrogel system. Physical structures of these nanofiber networks are quite similar to the structure of fibrous extracellular matrix (9, 10). Besides having similar structure to ECM, these scaffolds can be produced with similar bioactivity by incorporating short peptide sequences in the epitope region, which are derived from the cell surface receptor interacting domains of ECM proteins. RGDS (11), KRSR (12), YIGSR (13), DGEA (12), REDV (14), and IKVAV (9, 15) are some of these sequences that have been previously used to obtain bioactive peptide nanofiber scaffolds. IKVAV is a peptide sequence derived from laminin, and incorporation of this sequence in peptide nanofiber scaffolds has been shown to direct NSC differentiation into neurons rather than astrocytes or oligodendrocytes (15).

In this chapter, production of peptide nanofiber scaffolds from peptide amphiphile molecules, their characterization, and utilization for NSC culture and differentiation are described. Protocols for analyzing the effect of the scaffold on NSC viability and proliferation as well as analysis of the scaffold effect on differentiation are provided. Immunostaining of cell type-specific proteins along with qRT-PCR and western blot analysis is explained in detail. In addition, a method for quantitative analysis of morphological changes after differentiation is provided.

2 Materials

2.1 Cell Culture

1. Serological pipettes (Axygen), 15 and 50 mL conical tubes (CAPP), sterile pipette tips (filtered tips are preferred, Expel), tissue culture flasks and plates (Corning).

2. NSC culture medium: KnockOut DMEM/F-12 medium supplemented with 1 % GlutaMAX-I, 2 % StemPro NSC SFM Supplement, 20 ng/mL bFGF and EGF (all from Invitrogen). It is stable for 4 weeks when stored in the dark at 4 °C. Prepare 100 mL each time, and work in aliquots to avoid exposing it to 37 °C multiple times.

3. NSC differentiation medium: Same with NSC culture medium except that it is not supplemented with bFGF and EGF.

4. PBS without calcium and magnesium.

5. Accutase (Invitrogen).

2.2 Viability and Proliferation Tests

1. Alamar Blue reagent (Invitrogen).

2. Live/Dead Assay (Invitrogen).

3. Cell proliferation ELISA, Colorimetric BrdU assay (Roche).

2.3 Immunocyto-chemistry

1. 13 mm diameter, round glass coverslips (Thermo Scientific).

2. 4 % paraformaldehyde (PFA) solution (16): Add 4 g PFA (see Note 1), 4 g sucrose, 10 mL 10× PBS, 0.5 mL 1 M $MgCl_2$, 2 mL 0.5 M EGTA into 75 mL ddH_2O and mix on a magnetic stirrer. Add 10 N NaOH dropwise until solution clears. Then, adjust pH to 7.4 by dropwise addition of 6 N HCl, and adjust volume to 100 mL by adding ddH_2O. Aliquoted solution can be stored at −20 °C until use. Heat is not preferred for dissolving PFA since higher temperatures can lead to generation of formic acid which results in increased background staining. Instead, pH transition is used to dissolve PFA in this protocol. $MgCl_2$ along with EGTA is used to preserve cytoskeletal structure, and sucrose is used to maintain overall morphology during fixation.

3. 0.3 % TritonX solution: Dissolve TritonX detergent in PBS to a final concentration of 0.3 % (v/v) on a magnetic stirrer until

you obtain a homogeneous solution. Solution can be stored at room temperature (RT).

4. Blocking solution: Prepare by adding 1 % BSA (diluted from a 10 % stock in PBS) and 10 % normal goat serum into 0.3 % TritonX solution. This solution is prone to contamination; thus prepare it fresh. 10 % BSA solution along with goat serum are stored at −20 °C in 1 mL aliquots.

5. Primary antibody dilution solution: Prepare similarly to blocking solution except that final concentration of normal goat serum here is 3 %.

6. Secondary antibody dilution solution: Dilute 10 % BSA stock solution to a final concentration of 1 % in PBS.

7. TO-PRO-3 (Invitrogen) diluted solution: Dilute 1 mM stock solution to 1 μM final concentration in PBS.

8. ProLong Gold Antifade Mounting Medium (Invitrogen).

2.4 RNA Isolation and qRT-PCR

1. TRIzol (Invitrogen).

2. Chloroform.

3. Isopropanol.

4. Ethanol.

5. DNase/RNase free water (Gibco).

6. DNase/RNase-free microcentrifuge tubes and pipette tips.

7. Temperature-controlled centrifuge.

8. Primer sets for target genes and at least one housekeeping gene.

9. SuperScript®III Platinum® SYBR®Green One-Step qRT-PCR Kit (Invitrogen).

10. A thermal cycler with real time analysis property and software (BioRad).

2.5 Cell Lysate Preparation and Western Blot

1. Lysis buffer: Prepare a 4× stock without protease inhibitor (PI) cocktail. Dilute to 1× with ddH_2O, and add PI at a final concentration of 2 % immediately before use. Prepare 4× lysis buffer by mixing 2 mL of 1 M Tris–HCl, pH 6.8 (final concentration: 200 mM), 160 μL of 0.5 M EDTA, pH 8.0 (final concentration: 8 mM), 4 mL of 10 % SDS (final concentration: 4 %), 400 μL β-mercaptoethanol (final concentration: 4 %), and 3.2 mL of glycerol (final concentration: 32 %). Adjust volume to 10 mL by adding ddH_2O. Mix by vortexing. Aliquot the 4× lysis buffer, and store at −20 °C until use.

2. Novex bis–tris 4–12 % acrylamide gels (Invitrogen).

3. Reagents and protein standards for Bradford protein assay (BioRad).

4. PVDF membrane (Thermo Scientific).

5. Filter papers for blotting (BioRad).

6. Semi-dry blotting system (BioRad).

7. Anode I buffer: 0.3 M Tris–HCl (pH 10.4), 10 % methanol.

8. Anode II buffer: 25 mM Tris–HCl (pH 10.4), 10 % methanol.

9. Cathode buffer: 25 mM Tris–HCl (pH 9.4), 10 % methanol, 40 mM glycine

10. Blocking solution: Dissolve nonfat dry milk (5 wt%) in new TBS buffer, and add 0.1 % (v/v) Tween 20. Mix on a magnetic stirrer until a homogeneous solution is obtained.

11. Tris-buffered saline (TBS): Prepare 10× TBS by mixing 200 mL of 1 M Tris–HCl, pH 7.5 (final concentration: 200 mM) and 292.2 g NaCl (final concentration: 5 M) on a magnetic stirrer. Adjust volume to 1 L by adding ddH$_2$O. Dilute to 1× with ddH$_2$O. Both 10 and 1× buffers can be stored at room temperature.

12. TTBS: Prepare by mixing 100 mL 10× TBS, 0.5 mL Tween-20 (final concentration: 0.05 %), and 900 mL ddH$_2$O. It can be stored at room temperature.

13. New TBS: Prepare by mixing 10 mL of 1 M Tris–HCl, pH 7.5 (final concentration: 0.01 M), 30 mL of 5 M NaCl (final concentration: 0.15 M), and 960 mL of ddH$_2$O. It can be stored at room temperature.

14. Novex Chemiluminescent Substrate (Invitrogen).

15. BioRad Imaging Station.

16. UREA buffer: 0.1 M Na$_2$HPO$_4$, 0.01 M Tris–HCl, 8 M urea in ddH$_2$O, adjust pH to 8.0.

3 Methods

3.1 Peptide Synthesis, Purification, and Characterization

Peptides are synthesized by a standard Fmoc-protected solid-phase peptide synthesis method. Synthesis, purification by high-performance liquid chromatography (HPLC), and characterization methods were previously described in detail (8).

1. After reverse-phase-high-performance liquid chromatography (RP-HPLC) purification of the peptide, evaporate acetonitrile by rotary evaporation and freeze-dry the peptide to obtain it in powder form.

2. Dissolve the peptide in desired concentration in ddH$_2$O. Mix by vortexing, and sonicate in ultrasound water bath until it completely dissolves. Check the pH of the peptide solution by using pH paper. Adjust the pH to 7 with NaOH or HCl solutions.

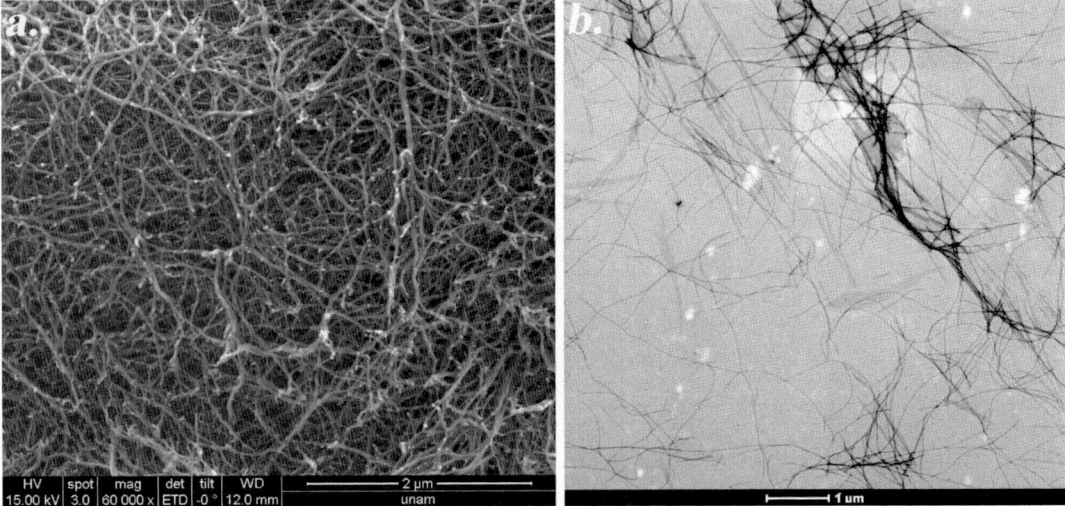

Fig. 1 (**a**) SEM image of the peptide nanofiber network, (**b**) TEM image showing individual peptide nanofibers

3. In order to analyze hydrogen bonding interactions, carry out circular dichroism (CD) measurements. Dissolve peptides to 2.5×10^{-4} M concentration as described above. Mix oppositely charged peptides in desired proportions, and mix by pipetting. Do not vortex in order not to disrupt the de novo-formed secondary structures. Detailed protocols are described here (8).

4. Stiffness of scaffold is an important physical property affecting the behavior of neural cells (17). In order to measure the stiffness of peptide nanofiber scaffolds, carry out rheological measurements. Prepare peptide gels of intended concentrations at a volume sufficient to fill the area of the upper plate. Prepare the gel on the lower plate, and incubate for at least for 10 min for gelation before measurement. Perform frequency sweep to determine storage modulus (G') and loss modulus (G''). The G' value being higher than G'' value indicates the gel behavior of the scaffold prepared. More detailed protocols for rheological analysis of peptide gels can be found here (8).

5. Characterize morphology of peptide nanofiber networks and peptide nanofibers individually by SEM and TEM imaging, respectively (Fig. 1). SEM imaging reveals the overall morphology of the scaffolds, while TEM is informative for investigating the morphology of individual fibers. It is possible to determine the diameter of peptide nanofibers from images taken by TEM. Detailed protocols of sample preparation and imaging parameters are explained here (8, 18).

3.2 Preparation of Scaffolds and NSC Culture

Peptide nanofiber scaffolds at physiological pH are produced by mixing oppositely charged peptide molecules dissolved in ddH$_2$O. Peptide nanofibers are formed by electrostatic interactions of oppositely charged side chains as well as hydrophobic interactions of the alkyl groups. By entrapment of water in between nanofiber networks during gelation, hydrogel scaffolds are produced.

1. Dissolve peptide powders in sterile ddH$_2$O at intended concentrations (usually 0.1–1 wt%, see Note 2). Adjust pH to 7.4 as explained in Section 3.1, and sonicate in ultrasound water bath until they completely dissolve. UV sterilize peptide solutions for at least 30 min.

2. For peptide nanofiber gel formation, add one of the oppositely charged peptide solutions to the well and then add the second solution via mixing with the pipette tip (with a swirling motion). A total volume of 300 µL is enough for 24-well plate wells. You can adjust the volume for larger or smaller surfaces accordingly. While preparing gels, consider the final charge of the scaffold and mix two peptides in proper ratios accordingly. An optimization for the charge might be required to obtain optimal bioactivity of the scaffold.

3. After coating of wells with peptide gels, incubate the plate at 37 °C for 30 min to 1 h for gelation to proceed and dry overnight under sterile conditions in a laminar flow hood. This step is carried out to evaporate water in the scaffold and obtain a nanofibrous network coating of the culture well surface. This network swells again with the culture medium and forms a very thin layer of gel beneath cells.

4. UV sterilize the peptide nanofiber-coated plate for at least 1 h.

5. Seed NSCs at a density of 5 × 10^4 cells/cm^2 in 500 µL/well for 24-well plate (see Note 3). You can adjust the volume for larger or smaller surfaces accordingly. After seeding, gently swirl the plate with your hand to obtain a homogeneous distribution of cells. Note that this protocol describes adherent culture of NSCs, not suspension culture as in the form of neurospheres. Incubate the plate at 37 °C, 5 % CO$_2$ in a humidified incubator. Change medium every 3 days.

3.3 Analysis of NSC Behavior on Scaffolds

3.3.1 Viability

Viability of the cells can be determined by either Alamar Blue Test or Live/Dead Assay as described below. Both detect live cells through metabolic activity of live cells, while Live/Dead Assay also detects dead cells by plasma membrane disintegrity. Detecting dead cells along with live cells is advantageous in that it directly provides data related to any cytotoxic effect leading to cell death. When the number of living cells is observed to be reduced by Alamar Blue, where only live cells are detected, it is not easy to understand if such a drop is caused by decreased cell proliferation or cell death. Detecting dead cells at the same time becomes advantageous in such a case.

Alamar Blue Test

Alamar Blue (AB) Test is a metabolic activity-based viability test. Active ingredient of AB reagent is resazurin, which is blue in color and non-fluorescent. Resazurin can readily enter into cells (it is cell permeable), and upon entering, it is reduced to resorufin by active metabolism of live cells. Resorufin is pink and fluorescent. The amount of resorufin produced and absorbance/fluorescence measured are directly proportional to the number of living cells.

1. Coat the wells of a 96-well plate with peptide nanofibers as described in Section 3.2. Prepare peptide nanofiber-coated wells to be used as blank as well. This is to avoid any background absorbance that might be caused by peptide nanofibers.

2. Seed cells at a density of 5×10^3 to 2×10^4 cells/well depending on the incubation time prior to viability measurement. Too low cell density leads to insufficient color development leading to unreliable results, while too high cell density leads to contact inhibition as well as cell death which again results in unreliable data. Add medium without cells to wells that will be used as blank.

3. At the time of analysis (usually in between 24 and 72 h), discard medium from wells and add 110 μL of fresh medium with Alamar Blue reagent (100 μL medium+10 μL AB/well). 10 μL of AB can also be added directly to medium over cells without discarding the medium, but in the case of long culture periods, non-even evaporation from wells might lead to some inconsistency. Hence, it is better to discard all medium and add same amount to all wells. This step should be carried out at dark (see Note 4).

4. Incubate the plate in a CO_2 incubator (37 °C, 5 % CO_2) for 2–4 h, depending on cell density. Color change from blue to pink can be tracked to determine the time of measurement.

5. Measure absorbance at 570 nm (reference λ: 600 nm). Alternatively, fluorescence can also be measured (excitation: 560 nm, emission: 590 nm).

Live–Dead Assay

Live–Dead Assay is a viability test which is used to detect live cells as well as dead cells. This assay is based on esterase activity of live cells and loss of plasma membrane integrity of dead cells. Calcein AM is a cell-permeant, non-fluorescent molecule which is cleaved by intracellular esterases upon entering cells. Calcein formed after esterase cleavage is well retained in live cells (it is cell impermeant) and gives an intense green fluorescence by which live cells can be detected. Dead cells are distinguished by ethidium homodimer staining. Ethidium is excluded by plasma membrane in live cells, while it is retained in dead cells due to disintegrity of plasma membrane. Dead cells can be detected by red fluorescence of ethidium homodimer dye.

1. Coat the wells of a 96-well plate with peptide nanofibers as described in Section 3.2.

2. Seed cells at a density of 5×10^3 to 2×10^4 cells/well depending on the time at which viability will be determined. Too low density leads to insufficient cell number in images leading to statistically unreliable results, while too high cell density leads to difficulties in distinguishing cell borders which again results in unreliable data.

3. At the time of analysis (usually in between 24 and 72 h), centrifuge the plate at $650 \times g$ for 5 min in a centrifuge with a plate rotor. This step is essential to detect dead cells that usually detach and float in the medium. If not centrifuged, they will be discarded in the next step and will not be detected.

4. During centrifugation, prepare calcein + ethidium working solution by diluting calcein and ethidium to a final concentration of 2 and 4 μM, respectively, in PBS. Since 200 μL/well will be used, prepare total volume according to the number of wells in the experiment. This step should be carried out in the dark to avoid fading of ethidium dye.

5. After centrifugation, discard medium over cells and add PBS slowly avoiding to detach cells. Carry out a brief (1–2 min) centrifugation at $650 \times g$. PBS wash is carried out to remove any serum esterase that cleaves calcein and leads to nonspecific background fluorescence.

6. After PBS wash, add 200 μL/well calcein+ethidium working solution (prepared in step 4) and incubate the plate for 30 min at room temperature (RT) in the dark.

7. After incubation, take merged images of calcein-stained live cells (green) and ethidium-stained dead cells (red) immediately by using an inverted fluorescent microscope. Take at least five random images/well.

8. Count the number of live and dead cells with Image J (Plugin: Cell Counter). Determine the % of live and dead cells on each substrate. Fluorescence of ethidium and calcein can also be measured alternatively as described in the product's manual. However, ethidium nonspecifically adheres to peptide nanofibers and produces high background due to which fluorescence measurement is not preferred. Dead cells are stained more intensely than the background and can be distinguished in images taken for counting.

3.3.2 Proliferation Test (BrdU Assay)

Stem cells, including NSCs, proliferate in culture. NSCs differentiate into neurons, astrocytes, or oligodendrocytes which are all nonproliferating cells. Hence, upon induction of differentiation a reduction in proliferation might indicate differentiation. However, this should not be used as the only data for verification of differentiation, since NSCs can also lose their proliferative ability due to cell senescence at later passage numbers. BrdU assay is commonly used for detection of proliferation. BrdU is a thymine analog, and it is incorporated into newly synthesized DNA in dividing cells; BrdU levels measured by ELISA are directly proportional to the number of proliferating cells.

1. Coat the wells of a 96-well plate with peptide nanofibers as described in Section 3.2. Prepare peptide nanofiber-coated wells to be used as blank as well. This is to avoid any background absorbance that might be caused by peptide nanofibers.

2. Seed cells at a density of 5×10^3 to 2×10^4 cells/well depending on the time at which proliferation will be determined. Too low density leads to insufficient color development leading to unreliable results, while too high cell density leads to contact inhibition as well as cell death which again results in unreliable data. Add medium without cells to wells that will be used as blank.

3. 2–24 h (depending on cell density and doubling time of cells) prior to the detection, discard medium over cells and add 100 μL of fresh medium with 10 μM BrdU reagent. 10 μL of BrdU from stock can also be added directly to medium over cells without discarding the medium, but in case of long culture periods, non-even evaporation from wells might lead to some inconsistency. Hence it is better to discard all medium and add same amount to all wells. Incubate cells under standard cell culture conditions until the analysis (2–24 h later).

4. Discard the medium over cells by tapping or suction. The assay can be interrupted at this step to continue later. Dried cells can be stored for up to 1 week at 2–8 °C until analysis.

5. Fix cells by adding 200 μL/well fixation solution provided in the kit, and incubate for 30 min at RT.

6. Remove fixation solution by flicking off and tapping. Add 100 μL/well peroxidase-labelled BrdU antibody working solution (1:100 diluted from stock), and incubate for 90 min at RT.

7. Discard antibody solution by flicking off, and rinse wells with 200 μL/well PBS three times.

8. Discard PBS by tapping, and add 100 μL/well peroxidase substrate. Incubate at RT until color development is sufficient for detection (5–30 min). Measure absorbance at 372 nm (reference λ: 492 nm). Repeated measurements can be done in order to determine the optimal time of measurement.

3.4 NSC Differentiation and Analysis of Differentiation

Forty-eight hours after seeding cells on peptide nanofiber scaffolds in NSC culture medium, discard the medium over cells and add NSC differentiation medium. Note that this medium is suitable for spontaneous differentiation of NSCs into neurons, astrocytes, and oligodendrocytes. Such an experiment is useful to determine the effect of the scaffold bioactivity in cell fate. If differentiation into a

Table 1
Markers commonly used for analysis of cell fate after NSC differentiation

Cell type	Markers	Reference
NSC	Nestin	(19)
Neuron	Map2, β-III-tubulin	(20–22)
Astrocyte	GFAP	(23)
Oligodendrocyte	GalC	(24)

specific cell type is desired, additional growth factors should be included in the differentiation medium.

3.4.1 Immunocyto-chemistry

Immunostaining against specific markers of NSCs, neurons, astrocytes, and oligodendrocytes is a powerful method in analysis of spontaneous differentiation as it allows to determine which cell fate is more dominantly directed by the scaffold. Most commonly used markers for this analysis are shown in Table 1.

1. For immunostaining, coat peptide nanofibers on 13 mm glass slides placed in 24-well plates, and culture and differentiate NSCs on these surfaces.

2. At the intended time point for analysis, discard the medium over cells and wash wells with PBS (750 μL/well). Be careful not to detach cells in this step.

3. Fix cells with prewarmed 4 % paraformaldehyde, 500 μL/well for 15 min at RT.

4. Permeabilize cells with 500 μL/well 0.3 % TritonX for 15 min at RT. Slowly mix on shaker.

5. Wash with 750 μL/well PBS for 5 min three times on shaker.

6. Add 500 μL/well blocking solution, and incubate for 1 h at RT. Slowly mix on shaker.

7. Add 100 μL/well diluted primary antibody (see Note 5 for co-staining), seal the plate with parafilm, and incubate overnight at 4 °C on a shaker at slow rate (a nutator/rocking shaker is preferred). Next day, continue with an additional 30-min to 1-h incubation at RT on shaker. For samples that will be used as negative control of staining, add primary antibody dilution solution described in Section 2.3 instead of primary antibody.

8. Wash with 750 μL/well 0.3 % TritonX for 5 min on shaker.

9. Wash with 750 μL/well PBS for 5 min on shaker.

10. Add 100 μL/well diluted secondary antibody that is conjugated with a fluorescent dye for detection. Incubate for 1 h at RT on a shaker at slow rate (a nutator/rocking shaker is preferred).

Starting from this step work in the dark in order to avoid fading of fluorescence.

11. For nuclear staining, add 300 μL/well TO-PRO-3 (1 μM final concentration) and incubate at RT for 15 min (see Note 6 for other nuclear staining options).

12. Wash with 750 μL/well PBS for 5 min three times on shaker.

13. Wash with 750 μL/well ddH$_2$O for 5 min on shaker. This step is essential to avoid salt crystals that form during drying due to the remaining PBS.

14. Label slides properly, and add one drop of ProLong Gold Antifade Mounting Medium onto each slide avoiding bubbles.

15. Remove cover slip from well, invert, and put on the mounting solution on the slide. Avoid making air bubbles during this step.

16. Let the cover slip dry, and seal the edges with nail polish. Slides can be stored in the dark at 4 °C for at least a couple of months.

17. Take images with a fluorescent/confocal laser scanning microscope.

3.4.2 RNA Isolation and qRT-PCR

All plastic materials and solutions used should be RNase free, and experiments should be carried out under class I laminar flow hood.

RNA Isolation

RNA isolation from low cell numbers is not efficient, so it is better to carry out differentiation in 6-well plates for this experiment. Use cold reagents (stored at 4 °C), and carry out centrifugation steps at 4 °C in order to avoid RNA degradation.

1. Discard medium from wells, and add 1 mL/well TRIzol. Carry out this step in a fume hood as TRIzol is toxic. Swirl the plate for efficient distribution of TRIzol through wells, and collect cell lysates in 1.5 mL labelled microcentrifuge tubes. After this step, samples can be stored at −80 °C until RNA isolation.

2. Add 200 μL/tube chloroform, shake tubes vigorously for 15 s, and incubate for 2–3 min at RT.

3. Centrifuge samples at 21,500 × g for 17 min. After centrifugation, three layers with pink, white, and clear colors appear from the bottom to top. Transfer 500 μL of the upper (clear) phase into a new tube for each sample trying not to disturb bottom layers.

4. Add 500 μL/tube isopropanol, and mix by inversion 15 times. Incubate for 10 min at RT.

5. Centrifuge samples at 6,100 × g for 12 min. Discard supernatant, and add 1 mL of 75 % EtOH. Do not dissolve the pellet; just detach it from the tube wall.

6. Centrifuge samples at $6{,}100 \times g$ for 8 min. Discard supernatant, and add 1 mL pure EtOH. Do not dissolve the pellet; just detach it from the tube wall.

7. Centrifuge samples at $6{,}100 \times g$ for 8 min. Discard supernatant, and air-dry pellets for 3–5 min. Do not overdry as it reduces solubility of RNA in water. Add 30 µL/tube DNase/RNase-free water, and dissolve the pellet by pipetting.

8. Measure RNA concentration and absorbance ratios of 260/280 and 260/230 nm at Nanodrop. 260/280 ratios of 1.9–2.1 and 260/230 ratios of 2–2.3 are acceptable for use of RNA in qRT-PCR. Isolated RNA should be stored at -80 °C until use.

qRT-PCR

For analysis of expression of marker genes, Pfaffl method is used for analysis of qRT-PCR results where primer efficiencies are also considered. We carry out cDNA synthesis and amplification of cDNAs in one reaction by using SuperScript®III Platinum® SYBR®GreenOne-Step qRT-PCR Kit (Invitrogen), which is a conventional and time saving method. The below protocol is for the use of this method.

1. Design primers specific for the target genes. Try to select primers from exon–exon boundaries so that the DNA cannot be amplified (due to the long intron sequence involved, which is not present in RNA). If not possible, DNase treatment should be carried out to avoid amplification of DNA besides RNA.

2. Before expression analysis, optimize annealing temperature for primers. After that, determine primer efficiency by carrying out qRT-PCR with serial dilutions of RNA. Plot a standard curve with the Ct values (y-axis) and RNA concentrations (x-axis), and determine the slope of the curve. Efficiency should be in the range of 90–110 % for reliable PCR results. Efficiency (E) can be calculated by using the formula $E = 10^{(-1/\text{slope})} - 1$.

3. Carry out qRT-PCR using primers for target genes as well as those for housekeeping genes. From Ct values obtained, perform a gene expression analysis by normalizing the results to those of housekeeping genes. You can now compare the expression between different experimental groups.

3.4.3 Cell Lysate Preparation and Western Blot

In order to obtain enough protein amount in cell lysate, using high cell numbers (at least 1.5–2×10^5 cells) is essential. Hence, it is better to carry out differentiation in 6-well plates for this experiment. During cell lysate preparation from cells cultured on uncoated surfaces, adding lysis buffer directly to the well and scraping is a conventional method. However, for peptide nanofiber-coated surfaces, this method is not suitable as peptide scaffold is also collected and contributes to the protein amount detected during protein assay. This leads to confusion in understanding the total protein amount obtained from cells. For this reason, prefer to detach cells enzymatically and lyse them later. For blotting, we prefer to use semidry blotting by using Bio-Rad's equipment, and the below protocol is optimized for this method.

1. Detach cells by using conventional trypsin detachment method. Since cells might adhere to peptide scaffolds very strongly, they might not detach with only trypsin treatment. If you have problems in detaching cells by this method, you can try 1:1 mixture of trypsin (0.25 % trypsin–EDTA):collagenase (1 %). Incubate cells with this mixture for 30 min at 37 °C, and observe every 10 min under microscope after tapping to help in detachment. Almost 90 % of cells detach with this method from the surfaces where cells adhere very strongly.

2. After detaching cells, centrifuge at $650 \times g$ for 5 min to obtain cell pellet. Dissolve the pellet in PBS, and centrifuge again.

3. Discard the PBS, and resuspend the pellet in 50–100 μL lysis buffer depending on the pellet size (see Section 2.5).

4. Incubate samples on ice for 30 min. Vortex every 10 min.

5. Sonicate samples on ice with a probe sonicator for 15–30 s.

6. Heat samples to 95 °C for 5 min, and cool on ice. This step is essential to denature DNA, which causes viscosity in samples. Samples can be stored at −80 °C after this step until use.

7. Determine the total protein amount in cell lysates by a method of your choice. We use Bradford Assay in our laboratory. Take care that the assay you choose does not interfere with the components of the lysis buffer and give false-positive results.

8. Load samples (30–100 μg) and molecular weight marker to 4–12 % commercial gels for SDS-PAGE, and run the gel according to the manual provided with the gel. In order to check that blotting occurs properly and your antibodies work perfectly, you can load a positive control to one of the wells. Brain lysate can be used for this purpose as it contains all the neural marker proteins (see Note 7 for preparation of brain lysate).

9. For blotting, cut the PVDF membrane and extra thick filter papers to the dimensions of the gel. Three pieces of filter papers are required for one gel.

10. Equilibrate the gel and membrane in ice-cold transfer buffers (see Section 2.5) as described in Table 2.

11. Assemble the blotting unit, and place the filter papers, gel, and membrane according to Table 2. The filter paper which was wet

Table 2
Equilibration of filter papers, gel, and membrane before blotting

Layer starting from the bottom	Material	Equilibration
1	Filter paper	Wet with anode buffer I
2	Filter paper	Wet with anode buffer II
3	PVDF membrane	15 min with methanol, 2 min with ddH$_2$O, 5 min with anode II on shaker
4	Gel	15 min with anode II
5	Filter paper	Wet with cathode buffer

with cathode buffer will be at the top. Roll out air bubbles at every layer by using a glass pipette.

12. Carry out blotting at 14 V for 30 min.

13. After blotting, place the membrane in blocking solution (see Section 2.5) and incubate for 2 h at RT, shaking slowly.

14. Discard blocking solution, and add primary antibody diluted in blocking solution. Incubate overnight at 4 °C, shaking slowly. Next day, incubate at RT for 30 min to 1 h.

15. Wash with blocking solution for 20 min at RT with shaking.

16. Add secondary antibody diluted in blocking solution. Incubate for 45 min at RT, shaking slowly.

17. Wash membrane with TTBS for 5 min twice and TBS for 5 min twice, with shaking.

18. Wash membrane with ddH$_2$O for 5 min.

19. Prepare substrate (Novex Chemiluminescent Substrate) immediately before use by mixing Reagent A and Reagent B in 1:1 ratio. It is essential to warm substrate reagents to RT before use. Work in a darkroom from this point forward.

20. Add substrate on the membrane, and incubate for 1 min. Use substrate volume enough to cover the surface of the membrane.

21. Image the membrane with BioRad imaging station, VersaDoc software by selecting Blotting, "Chemiluminescence, Ultra-High Sensitivity" from the software with 600 s exposure time.

3.4.4 Analysis of Neurite Outgrowth

Neurite outgrowth is an indication of neuron's response to axon guidance cues provided by the environment and hence an important indication of the bioactivity of the scaffold used in directing NSCs towards differentiation. Analysis of neurite outgrowth by ImageJ program is described in this section. ImageJ is an NIH-released free program that can be downloaded from http://rsbweb.nih.gov/ij/.

1. At the intended time of analysis (multiple time points during differentiation is useful), take at least five random images from each sample. You should study with at least three replicates (n) within each experimental group.

2. Open the image to be analyzed with ImageJ program. Select segmented line in the tabs which allows to track neurites even when they are curved. From shortcuts list, find the shortcut for "Measure and Label." This shortcut is a letter in the keyboard which you can use to measure the length of the neurite you tracked. Besides it labels the track so that you do not measure it twice. After labeling and measuring all neurites in the image you can transfer results to Excel for further analysis such as the total neurite length in the image. Besides, by using the cell counter plug-in, you can count the total number of cells in the image as well as the cells extending neurites.

3. After measuring the total neurite length, total number of cells, and number of cells extending neurites, you can now calculate the average neurite length in that image as well as % of cells extending neurites. Note that the length measured in ImageJ is in arbitrary unit (au). You can convert this to μm by measuring the scale bar in the image and finding the conversion factor between au and μm. By using this conversion factor, you can now express the average neurite length in the image in μm scale.

4. After completing the measurement of all images, take average of random images from the same well and after that take average of different samples of the same experimental group and find the standard deviation within the group.

4 Notes

1. PFA is toxic; weigh and prepare the solution in a fume hood.

2. Gelation properties and bioactivity of different peptides might differ. Hence, it is better to carry out an optimization with different concentrations of peptides for scaffold preparation. Lower concentrations might not lead to gel formation, while too high concentrations result in opaque gels leading to difficulty in observing cells under optical microscope. Besides, the concentration with optimal bioactivity should be determined with such an optimization assay.

3. We use rat fetal NSCs obtained from Invitrogen. These cells are isolated from cortex of embryonic day 14 Sprague–Dawley rats. Seeding density of 5×10^4 cells/cm^2 is suggested for

these cells to avoid spontaneous differentiation. If using cells from other sources, it is better to optimize seeding density prior to other experiments.

4. Avoid exposing Alamar Blue reagent to light as it is reduced in light which leads to false-positive results.

5. Co-staining of different proteins in same samples is a powerful method for determining % of different cell types in the same population of cells. In order to carry out co-staining, you need to select primary antibodies from different sources (such as one from mouse, the other from rabbit) and secondary antibodies specific to these primary antibodies should be labelled with different fluorophores. In this case, you can incubate same samples with both primary antibodies at the same time. In the secondary antibody treatment, you should also add both secondary antibodies to the same well. While imaging, you can take merged images showing expression of both proteins.

6. You can also use DAPI or some other fluorescent dyes commercially available for nuclear staining. However, check whether the excitation and emission wavelengths are suitable with your fluorescent/confocal microscope before carrying out staining.

7. Brain lysate preparation: We prepare lysate from frozen (stored at -80 °C) mouse brain. It can also be prepared from fresh tissue. Mince brain with scalpel after weighing. Add 1 mL of urea buffer to 1 g tissue, and sonicate with a probe sonicator on ice. Incubate at 95–100 °C for 5–10 min. Determine the total protein concentration by a method of your choice (we use Bradford method).

Acknowledgments

B.M. is supported by Scientific and Technological Research Council of Turkey (TUBITAK) grant number 111M410. M.O.G and A.B.. T. acknowledge support from the Turkish Academy of Sciences Distinguished Young Scientist Award (TUBA-GEBIP).

References

1. Temple S (2001) The development of neural stem cells. Nature 414(6859):112–117
2. Einstein O, Ben-Hur T (2008) The changing face of neural stem cell therapy in neurologic diseases. Arch Neurol 65(4):452–456
3. Yuan T et al (2013) Human induced pluripotent stem cell-derived neural stem cells survive, migrate, differentiate, and improve neurological function in a rat model of middle cerebral artery occlusion. Stem Cell Res Ther 4(3):73
4. Falk A et al (2012) Capture of neuroepithelial-like stem cells from pluripotent stem cells provides a versatile system for in vitro production of human neurons. PLOS One 7(1)

5. Thier M et al (2012) Direct conversion of fibroblasts into stably expandable neural stem cells. Cell Stem Cell 10(4):473–479

6. Ring K et al (2012) Direct reprogramming of mouse and human fibroblasts into multipotent neural stem cells with a single factor. Cell Stem Cell 11(1):100–109

7. Sharma K, Selzer M, Li S (2012) Scar-mediated inhibition and CSPG receptors in the CNS. Exp Neurol 237(2):370–378

8. Toksoz S, Mammadov R, Tekinay A, Guler M (2011) Electrostatic effects on nanofiber formation of self-assembling peptide amphiphiles. J Colloid Interface Sci 356(1):131–137

9. Mammadov B, Mammadov R, Guler M, Tekinay A (2012) Cooperative effect of heparan sulfate and laminin mimetic peptide nanofibers on the promotion of neurite outgrowth. Acta Biomater 8(6):2077–2086

10. Nishida T, Yasumoto K, Otori T, Desaki J (1988) The network structure of corneal fibroblasts in the rat as revealed by scanning electron-microscopy. Invest Ophthalmol Vis Sci 29(12):1887–1890

11. Guler M et al (2006) Presentation of RGDS epitopes on self-assembled nanofibers of branched peptide amphiphiles. Biomacromolecules 7(6):1855–1863

12. Anderson J et al (2009) Osteogenic differentiation of human mesenchymal stem cells directed by extracellular matrix-mimicking ligands in a biomimetic self-assembled peptide amphiphile nanomatrix. Biomacromolecules 10(10):2935–2944

13. Sur S et al (2012) A hybrid nanofiber matrix to control the survival and maturation of brain neurons. Biomaterials 33(2):545–555

14. Ceylan H, Tekinay A, Guler M (2011) Selective adhesion and growth of vascular endothelial cells on bioactive peptide nanofiber functionalized stainless steel surface. Biomaterials 32(34):8797–8805

15. Silva G et al (2004) Selective differentiation of neural progenitor cells by high-epitope density nanofibers. Science 303(5662):1352–1355

16. Marchenko S, Flanagan L (2007) Immunocytochemistry: human neural stem cells. J Vis Exp (7):267

17. Sur S, Newcomb CJ, Webber MJ, Stupp SI (2013) Tuning supramolecular mechanics to guide neuron development. Biomaterials 34 (20):4749–4757

18. Mammadov R, Tekinay A, Dana A, Guler M (2012) Microscopic characterization of peptide nanostructures. Micron 43(2–3):69–84

19. Lendahl U, Zimmerman LB, McKay RD (1990) CNS stem cells express a new class of intermediate filament protein. Cell 60 (4):585–595

20. Caccamo D et al (1989) Immunohistochemistry of a spontaneous murine ovarian teratoma with neuroepithelial differentiation. Neuron-associated beta-tubulin as a marker for primitive neuroepithelium. Lab Invest 60 (3):390–398

21. Matus A (1988) Microtubule-associated proteins: their potential role in determining neuronal morphology. Annu Rev Neurosci 11:29–44

22. Svendsen CN, Bhattacharyya A, Tai YT (2001) Neurons from stem cells: preventing an identity crisis. Nat Rev Neurosci 2(11):831–834

23. Pixley SK, Kobayashi Y, de Vellis J (1984) Monoclonal antibody to intermediate filament proteins in astrocytes. J Neurosci Res 12 (4):525–541

24. Rostami A et al (1984) Generation and biological properties of a monoclonal antibody to galactocerebroside. Brain Res 298(2):203–208

Methods in Molecular Biology (2014) 1202: 149–160
DOI 10.1007/7651_2013_34
© Springer Science+Business Media New York 2013
Published online: 24 October 2013

The Delivery and Evaluation of RNAi Therapeutics for Heterotopic Ossification Pathologies

Arun R. Shrivats and Jeffrey O. Hollinger

Abstract

RNA interference (RNAi) is a powerful tool being used to develop therapies for pathologies caused by gene overexpression. Heterotopic ossification pathologies such as trauma-induced heterotopic ossification and fibrodysplasia ossificans progressiva may be treatable with an RNAi approach. However, there is a lack of consensus in literature regarding the delivery conditions and evaluation of RNAi therapeutics in these disease models. Here, we describe in vitro protocols for the delivery of polymer-based RNAi therapeutics as well as a streamlined strategy for the assessment of osteoblast lineage progression due to dysregulated bone morphogenetic protein signaling. This strategy focuses on the quantification of early-stage osteoblast transcription factors RUNX2 and OSX, followed by the measurement of alkaline phosphatase activity and late-stage matrix deposition.

Keywords: RNA interference, siRNA delivery, Nanostructured polymer, Osteoblast lineage progression, Heterotopic ossification, Fibrodysplasia ossificans progressiva, Osteogenic differentiation, Osteoblastogenesis

1 Introduction

RNA interference (RNAi) is a powerful technique capable of achieving targeted gene silencing. Theoretically, RNAi machinery can be employed to knock down expression of any disease-causing gene, giving it a broad scope as a potential therapeutic. One area where this ability can be harnessed is the development of therapies for heterotopic ossification (HO) pathologies where extra-skeletal bone formation processes occur, such as trauma-induced HO and fibrodysplasia ossificans progressiva (FOP). In these pathologies, interference in traditional bone morphogenesis signaling pathways may be able to prevent the progressive heterotopic ossification with minimal off-target effects. While this line of research is promising, we require robust in vitro and in vivo analytic techniques to determine the ability of a therapeutic to disrupt abnormal signaling processes and impede osteoblast lineage progression.

Traditional in vitro analyses for HO and FOP treatments involve pre-osteoblast cell lines such as murine calvarial cells (MC3T3s) or murine myoblasts (C2C12s) and subsequently expose them to differentiation cues such as bone morphogenetic proteins or osteogenic media supplements (1–3). RNAi therapeutics are evaluated for their ability to impede progression of osteoblastic differentiation. The rationale behind such an approach is that these cell phenotypes may contribute to unwanted bone morphogenesis within soft tissues. The use of established cell lines in these studies provides populations of homogenous cells whose characteristics may be expected to be consistent and predictable, minimizing variation between studies. The trade-off of using cell lines may include difficulties in accurately representing the entire range of cell phenotypes that play a role in the pathophysiology of HO pathologies. An alternative to cell lines is primary cells, which may provide a more representative in vitro model. Primary osteoblasts may be harvested from the calvaria of mice (or other species) for studies of osteoblast differentiation (4). While the homogeneity of the cell population cannot be absolutely guaranteed, there are benefits that may offset this. By harnessing our knowledge of genetics, transgenic animal models have been developed that mimic human HO pathologies (5). One such example is the introduction of Q233D, Q207D, and R206H mutations in mice to produce species that are hyperactive upon BMP ligand binding (4, 6). These are employed as cellular models for the evaluation of treatments for heterotopic ossification and FOP pathologies (4). Harvesting cells from these animal models can provide a more representative in vitro testing platform for therapeutics to help the transition to in vivo studies.

Whatever the origin of cellular models employed for the testing and development of therapeutics for HO pathologies, we require a comprehensive set of in vitro assays to reliably evaluate osteoblast lineage progression.

The delivery of rhBMP-2 to osteoprogenitor cells results in the expression of differentiation markers in predictable temporal patterns (7–15). Figure 1 illustrates the approximate expression profiles for several well-characterized osteoblast markers, guiding our in vitro analyses.

In order to determine osteoblast lineage progression, we must exploit our understanding of osteoblast marker expression and the subsequent changes in cell phenotype. Pivotal early markers in the differentiation process include runt-related transcription factor 2 (RUNX2), osterix (OSX), and alkaline phosphatase (ALP). These markers are expressed at predictable temporal periods after rhBMP-2 delivery. In later stages of osteoblast differentiation, while markers such as osteocalcin, osteonectin, and bone sialoprotein are

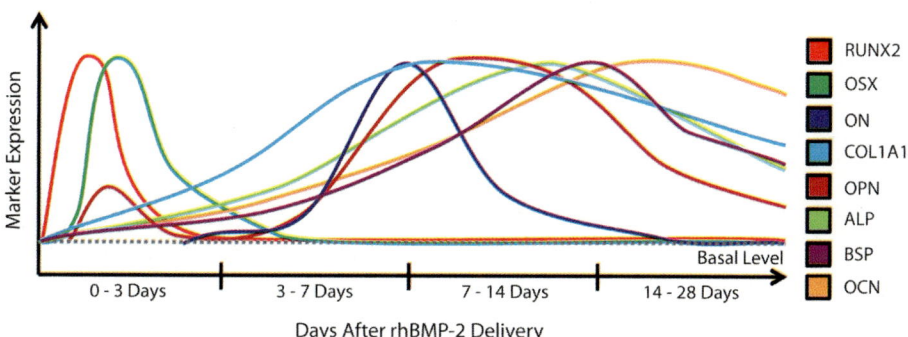

Fig. 1 Approximate temporal expression profiles of prominent osteoblast differentiation markers in pre-osteoblast cells after rhBMP-2 treatment. Marker legend is organized chronologically by peak expression level. Markers that peak within the first 3 days include runt-related transcription factor 2 (RUNX2) and osterix (OSX). Type I collagen (COL1A1) and alkaline phosphatase (ALP) are both produced within 48 h of rhBMP-2 delivery, but expression levels peak later in the differentiation process. Notable downstream markers include osteonectin (ON), osteopontin (OPN), bone sialoprotein (BSP), and osteocalcin (OCN)

present in detectable amounts, there are phenotypic differences in cell cultures that serve as more reliable indicators of differentiation. Matrix deposition and mineralization, for instance, are hallmarks of late-stage osteoblast differentiation and are effective methods for assessing the osteoblast differentiation endpoint. Thus, our strategy for the evaluation of osteogenic differentiation is as follows: At early stages of osteoblast differentiation, we focus on quantifying RUNX2, OSX, and ALP expression while at late stages, we evaluate cells for the presence of mineralization.

In this chapter, we describe a panel of finely tuned assays for the thorough evaluation of RNAi therapeutics in preventing osteogenic differentiation. However, given the lack of consensus in literature regarding the conditions required for successful RNAi therapeutics, a description of these techniques would be incomplete without including the prior steps that necessitate these analyses. As such, we begin with the delivery of siRNA as well as rhBMP-2-based induction of osteoblastic differentiation. We focus on quantification of differentiation in three temporal windows following a differentiation trigger: early stage (24–72 h), medium stage (4–14 days), and late stage (21–28 days).

2   Materials

2.1   Instruments

1. Horizontal gel electrophoresis apparatus with PowerPac™.

   (a) Including gel chamber, casting tray, and well combs.

2. AlphaImager (or any UV 265 nm transilluminator with a camera).

3. ABI Prism® Sequence Detection System (7000+).

4. pH meter.

5. Multi-plate shaker.

6. Tecan Spectra Fluor (or any fluorescence and absorbance reader).

2.2 Cell Culture Supplies

1. Complete α-MEM media.

   (a) 10 % fetal bovine serum (FBS) (Invitrogen, 10082-139).

   (b) 1 % penicillin/streptomycin (Pen–Strep) (Invitrogen, 15140-122).

   (c) 89 % incomplete α-MEM media (Invitrogen, A10490-01).

2. Recombinant human bone morphogenetic protein 2 (rhBMP-2).

3. Opti-MEM® (Life Technologies, 11058-021).

4. Anti-*Runx2* and anti-*Osx* siRNA with delivery vehicle.

5. Nuclease-free water (Invitrogen, AM9930).

6. Deionized (DI) water.

7. 70 % ethyl alcohol (Sigma-Aldrich, E7023).

2.3 Assay Reagents

Gel Electrophoresis

1. 1× TBE buffer

   (a) Trizma base (Sigma, T1503).

   (b) Boric acid (Sigma, B7901).

   (c) 0.5 M EDTA (Invitrogen, 15575-020).

2. Agarose (Sigma-Aldrich, A4718).

3. Ethidium bromide (Sigma, E1510).

Quantitative Reverse Transcriptase PCR

4. CellsDirectTM One-Step qRT-PCR Kit with ROX (Invitrogen 11754-100).

5. Gene-specific predesigned primers compatible with the Taqman® Gene Expression Assay—obtained from Life Technologies.

   (a) RUNX2 (Mm00501580_m1).

   (b) OSX (Mm04209856_m1).

   (c) ACTB (Mm00607939_s1).

   (d) HMBS (Mm01143545_m1).

Alkaline Phosphatase Activity

6. Cell Lysis Buffer (Cell Signaling Technology, #9803).

7. Alkaline working buffer (Sigma, A9226).

8. 5 mM *p*-nitrophenyl phosphate substrate (PNPP) (Sigma, N2770).

9. 10 mM *p*-nitrophenyl standard solution (Sigma, N7660).

10. 1 N NaOH (Sigma, 319511).

11. Protein Assay Kit (Bio-Rad, 500–0001).

12. Quant-iT™ PicoGreen® dsDNA Assay Kit (Life Technologies, P11496).

Mineralization Studies

13. 5 % silver nitrate solution (Sigma-Aldrich, S6506).

14. 4 % formaldehyde (diluted from 16 % formaldehyde).

15. Alizarin Red S (Sigma-Aldrich, A5533).

16. Cetylpyridinium Chloride (CPC) (Sigma-Aldrich, C-0732).

17. Sodium phosphate monobasic (Sigma-Aldrich, S8282).

18. Sodium phosphate dibasic (Sigma-Aldrich, S7907).

19. 70 % ethyl alcohol (Sigma-Aldrich, E7023).

20. Sterile PBS (Invitrogen, 10010-023).

21. 0.1 N NaOH (Fluka, 319481).

22. 0.1 N HCl (Fluka, 318965).

3   Methods

3.1   Polymer:siRNA Complexation and Validation by Gel Electrophoresis

1. To prepare siRNA treatments, we begin with our nanostructured polymers (16, 17) and siRNA, both suspended in nuclease-free water (*see* **Note 1**).

2. Sequentially combine the solvent (nuclease-free water, if necessary to achieve a desired final volume), the nanostructured polymers, and the siRNA solutions in the appropriate volumes to prepare polymer-siRNA complexes.

   (a) Allow the polymers and siRNAs to complex at 4 °C for 1 h.

3. Upon being brought to room temperature, the treatments are ready for delivery to cells. If an evaluation of complexation is desired, employ gel electrophoresis (procedure continued below) to separate unbound siRNA.

4. Add glycerol (10 % by volume) to each polymer-siRNA sample to serve as a loading buffer (*see* **Note 2**).

5. Prepare 1× TBE buffer by combining the following ingredients: 10.6 g Trizma base, 5.5 g boric acid, 4 mL 0.5 M EDTA, and 1 L DI water. Mix using a magnetic stir bar until all components have dissolved.

6. Prepare 2 % (w/v) agarose gels with 1× TBE buffer, and load samples into the gel (*see* **Note 3**).

   (a) 1× TBE buffer is used as both the gel solvent and running buffer.

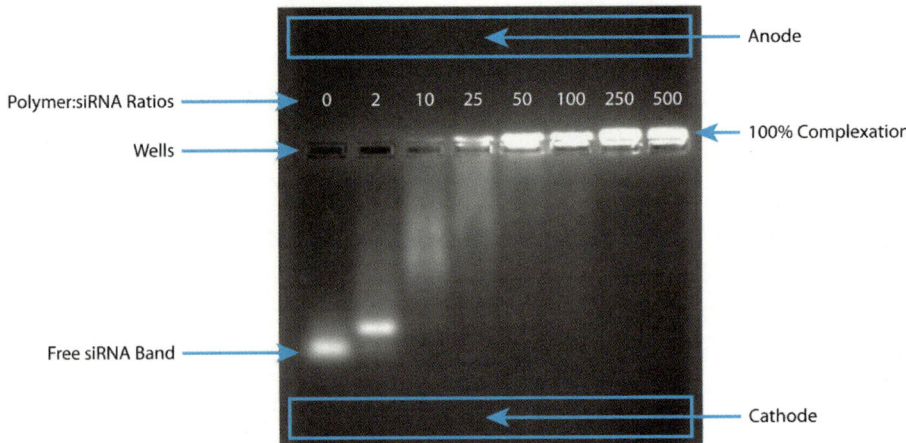

Fig. 2 Typical results of gel electrophoresis to evaluate complexation at various polymer:siRNA (weight:weight) ratios. Here we prepare polymer:siRNA solutions at ratios ranging from 0:1 to 500:1. Full (100 %) complexation is achieved when there is no evidence of siRNA migration from the well towards the cathode. In this case, this occurs between ratios of 100:1 and 250:1. The migration of the 0:1 sample demonstrates expected behavior of the free siRNA sample through the gel

7. Run the gel at 100 V for 30 min. To stain, submerge the gel in a solution of $1\times$ TBE buffer with 0.5 µg/mL ethidium bromide for 30 min, away from light.

   (a) Note: Ethidium bromide is a mutagen, and the appropriate laboratory attire should be used when handling it.

8. Image the gel under UV 265 nm transillumination. The results obtained will resemble Fig. 2.

3.2 siRNA Delivery Protocol

1. Seed cells 1 day prior to rhBMP-2 delivery to begin differentiation processes. We typically adjust the concentration of cells based on the cell phenotype and typical doubling time. Deliver rhBMP-2 when cells are 75–80 % confluent.

2. Deliver RNAi therapeutic. siRNA doses typically range from 10 to 30 pmol for cells seeded in 24-well plates (18, 19) (*see* **Note 4**).

3. In serum-free studies (*see* **Note 5**), wash cells prior to siRNA delivery with serum-free α-MEM in order to remove any serum components left in the samples. Replace growth culture medium with Opti-MEM® or serum-free α-MEM. After this media change, deliver siRNA treatments. After incubation at 37 °C for 4 h (5 % CO_2), add FBS to achieve a final FBS concentration in media of 10 %.

4. In serum-containing studies (*see* **Note 5**), deliver treatments immediately in medium supplemented with 10 % FBS and 1 % Pen–Strep and incubated at 37 °C (5 % CO_2).

3.3 Analysis of Short-Term Osteoblast Differentiation Factors

Quantification of short-term osteoblast differentiation markers RUNX2 and OSX can be achieved using quantitative real-time PCR (qRT-PCR). qRT-PCR allows for the simultaneous quantification of mRNA expression for multiple genes in numerous samples. To assess early osteoblast differentiation, we choose to investigate the expression of *RUNX2* and *OSX* genes encoding two transcription factors that are master regulators of osteoblastic differentiation (20, 21). Primers for target and reference genes are obtained from Invitrogen (*see* **Note 6**).

The protocol is as follows:

1. Perform cell lysis on cells 48 h after rhBMP-2 delivery according to the protocol from CellsDirect™.

2. Perform qRT-PCR with the following amplification conditions: 50 °C for 15 min, 95 °C for 2 min, and 40 cycles of 95 °C for 0.25 min and 60 °C for 1 min.

3. Calculate relative expression of target genes based on the ΔC_T method detailed by Schmittgen et al. (22) (*see* **Note 7**).

3.4 Osteoblast Differentiation Determination by Alkaline Phosphatase Activity

Alkaline phosphatase is an osteoblast differentiation factor that peaks 7–14 days into differentiation. For this purpose, we typically analyze expression at 4, 7, and 14 days following BMP-2 delivery.

The protocol is as follows:

1. Obtain cell lysates as per the CST lysis protocol. Briefly, prepare 1× lysis buffer by dilution with DI water. After washing cells with PBS, treat cells with 1× lysis buffer and incubate on ice for 5 min. Harvest cells using a cell scraper, and centrifuge contents at $14{,}000 \times g$ for 10 min. Transfer supernatants, and store at −20 °C when not using them.

2. Prepare the working solutions of each of the following reagents:

   (a) PNPP substrate solution: Reconstitute tablets in 20 mL of DI water (*see* **Note 8**).

   (b) 0.02 N NaOH: Dilute 100 μL of 1N NaOH with 49.9 mL DI water.

   (c) 0.3 N NaOH (stop solution): Dilute 15 mL of 1 N NaOH with 35 mL DI water.

   (d) ALP standards: Add 40 μL *p*-nitrophenyl standard solution to 1.96 mL of 0.02 N NaOH. Perform five serial dilutions in 0.02 N NaOH to produce solutions at concentrations of 200, 100, 50, 25, 12.5, 6.25, and 0 nmol/mL.

3. Bring all reagents to room temperature before use.

4. In a 96-well plate, add 80 μL of each lysate/standard solution to 20 μL alkaline working buffer.

5. Add 100 μL of PNPP substrate solution, and incubate at 37 °C for 10 h.

6. Add 100 μL of 0.3 N NaOH stop solution to each well to stop the reaction. Using a microplate reader, measure the optical density (absorbance) of samples at 405 nm.

7. Normalize ALP activity to the relative number of cells in each sample to obtain accurate assessment of ALP activity (*see* **Note 9**).

3.5 Osteoblast Differentiation Determination by Mineralization Staining

The von Kossa stain allows for the indirect detection of calcium phosphates. It involves the reaction of silver cations with any anions present in the samples (including phosphates, sulfates, and carbonates). This reaction results in the formation of a UV-photodegradable precipitate that allows for the determination of the presence of these phosphate groups. However, the nonspecificity of the von Kossa stain necessitates the confirmation of mineralization using an additional method as well (23). For this purpose, we employ alizarin red staining. At an alkaline pH, alizarin red S reacts with calcium ions to form soluble salts that are deep red in color. By utilizing both von Kossa and alizarin red staining, we safeguard ourselves against false positives among mineralization results (Fig. 3).

von Kossa Staining Protocol

1. Prepare 5 % (w/v) silver nitrate solution by dissolving 5 g of silver nitrate in 100 mL of deionized distilled water (*see* **Note 10**).

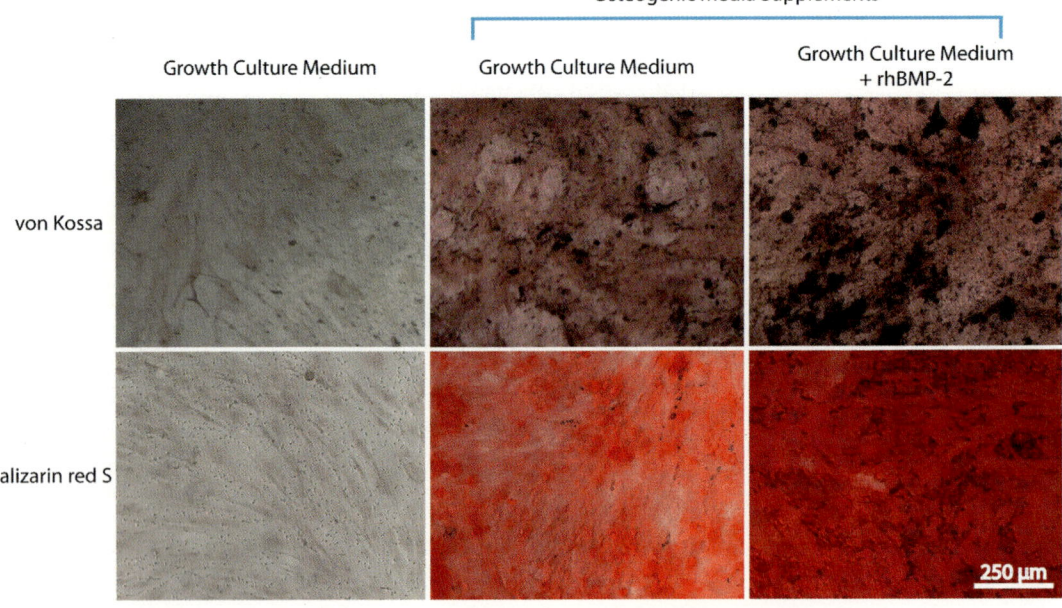

Fig. 3 An example of von Kossa and alizarin red staining on human mesenchymal stem cells cultured for 28 days in growth medium (*column 1*), growth medium with osteogenic supplements (*column 2*), and growth medium with osteogenic supplements and rhBMP-2 (*column 3*). The presence of rhBMP-2 in addition to osteogenic supplements has a noticeable impact on the intensity of mineralization

2. Remove cell culture media from cells, and rinse cells twice with 1× PBS. After the second rinse, add 1 mL of 4 % formaldehyde for 10 min.

3. Place under UV 265 nm irradiation for 30 min without the lid.

4. Rinse each well three times with DI water before adding 1 mL of 5 % sodium thiosulfate for 2–4 min (*see* **Note 11**).

5. Replace sodium thiosulfate with DI water following one final DI water rinse.

6. Leave samples in DI water, and image using bright field microscopy.

Alizarin Red Staining Protocol

1. Prepare working solutions of each of the following reagents:

   (a) 40 mM Alizarin red S: Add 6.85 g alizarin red S powder to 500 mL DI water, and stir for 5 min. Adjust pH to 4.2 using 0.1 N NaOH.

   (b) 0.2 M Monobasic sodium phosphate: Add 27.6 g NaH_2PO_4-H_2O in 1 L DI water.

   (c) 0.2 M Dibasic sodium phosphate: Add 53.65 g Na_2HPO_4-$7H_2O$ in 1 L DI water.

   (d) 10 % CPC: Prepare combined sodium phosphate solution by adding 39 mL monobasic sodium phosphate to 61 mL dibasic sodium phosphate. Dissolve 10 g CPC in 100 mL combined sodium phosphate buffer, and adjust pH to 7.0 using 0.1 N HCl and/or 0.1 N NaOH.

2. Aspirate cell culture medium from cells, and after washing with 1× PBS, add ice-cold 70 % ethanol to fix cells.

3. After 1 h, wash cells with DI water, and add 1 mL (per 24-well plate) of 40 mM alizarin red S. Incubate at room temperature for 10 min on a multi-plate shaker.

4. After 10 min, rinse cells three times with DI water, and add 1× PBS for 15 min (on a multi-plate shaker) to further reduce nonspecific staining.

5. Remove PBS, and image cells using bright field microscopy.

4   Notes

1. For testing in animals, high siRNA doses may be required with minimal therapeutic volumes. In order to prepare such polyplex solutions, we use polymers suspended at 40 μg/mL and siRNA at 200 μM.

2. The total volume of each sample should be between 25 and 50 μL. Each sample should contain 300–500 ng siRNA in order to achieve a clear siRNA signal after staining the gel.

3. We recommended the inclusion of an siRNA-only control.

4. When targeting the *Runx2* gene, deliver polymer–siRNA treatments 18–24 h prior to rhBMP-2 stimulation. With *Osx*, deliver treatments at the same time as rhBMP-2 stimulation.

5. We typically investigate siRNA delivery in two environments (sequentially): serum-free cell culture media followed by media with 10 % FBS. The purpose of serum-free siRNA delivery is to determine if knockdown of target gene expression is attainable under idealized delivery conditions with minimal interference to both the delivery mechanisms and the siRNAs themselves. This is done as a precursor to serum-containing environments; in these studies, we seek to deliver siRNAs in an in vitro environment that presented additional barriers to successful siRNA delivery (such as polyanionic competition and the presence of nucleases).

6. Successful execution of quantitative real-time PCR requires the selection of appropriate housekeeping genes. Popular choices are *18S* and/or *GAPDH*, though particularly with osteoblast differentiation studies, their use is controversial (24). An ideal housekeeping gene will be consistent over the course of the study and provide an accurate reflection of variations in cell population among different treatments. Based on several optimization studies published in literature, we recommend the use of the geometric mean of *ACTB* and *HMBS* to serve as a reference in osteoblast differentiation studies (25).

7. The $\Delta\Delta C_T$ method may be applied instead for analysis of qRT-PCR results, depending on experiment setup.

8. This solution can be stored at −20 °C for future use as well.

9. Normalization can be achieved by determining total protein or DNA content in cell lysates using either the Bradford protein or PicoGreen assays.

10. This solution can remain at room temperature for future use.

11. The addition of 5 % sodium sulfate minimizes false-positive results.

Acknowledgement

This work was supported by Department of Defense grant W81XWH-11-2-0073, through the Defense Medical Research and Development Program.

References

1. Pohl F, Hassel S, Nohe A, Flentje M, Knaus P, Sebald W, Koelbl O (2003) Radiation-induced suppression of the Bmp2 signal transduction pathway in the pluripotent mesenchymal cell line C2C12: an in vitro model for prevention of heterotopic ossification by radiotherapy. Radiat Res 159(3):345–350

2. Lin L, Chen L, Wang H, Wei X, Fu X, Zhang J, Ma K, Zhou C, Yu C (2006) Adenovirus-mediated transfer of siRNA against Runx2/Cbfa1 inhibits the formation of heterotopic ossification in animal model. Biochem Biophys Res Commun 349(2):564–572. doi:10.1016/j.bbrc.2006.08.089

3. Sakurai T, Sawada Y, Yoshimoto M, Kawai M, Miyakoshi J (2007) Radiation-induced reduction of osteoblast differentiation in C2C12 cells. J Radiat Res 48(6):515–521

4. Yu PB, Deng DY, Lai CS, Hong CC, Cuny GD, Bouxsein ML, Hong DW, McManus PM, Katagiri T, Sachidanandan C, Kamiya N, Fukuda T, Mishina Y, Peterson RT, Bloch KD (2008) BMP type I receptor inhibition reduces heterotopic [corrected] ossification. Nat Med 14(12):1363–1369. doi:10.1038/nm.1888

5. Fukuda T, Scott G, Komatsu Y, Araya R, Kawano M, Ray MK, Yamada M, Mishina Y (2006) Generation of a mouse with conditionally activated signaling through the BMP receptor, ALK2. Genesis 44(4):159–167. doi:10.1002/dvg.20201

6. Chakkalakal SA, Zhang D, Culbert AL, Convente MR, Caron RJ, Wright AC, Maidment AD, Kaplan FS, Shore EM (2012) An Acvr1 R206H knock-in mouse has fibrodysplasia ossificans progressiva. J Bone Miner Res 27(8):1746–1756. doi:10.1002/jbmr.1637

7. Huang W, Carlsen B, Rudkin G, Berry M, Ishida K, Yamaguchi DT, Miller TA (2004) Osteopontin is a negative regulator of proliferation and differentiation in MC3T3-E1 pre-osteoblastic cells. Bone 34(5):799–808. doi:10.1016/j.bone.2003.11.027

8. Hassan MQ, Javed A, Morasso MI, Karlin J, Montecino M, van Wijnen AJ, Stein GS, Stein JL, Lian JB (2004) Dlx3 transcriptional regulation of osteoblast differentiation: temporal recruitment of Msx2, Dlx3, and Dlx5 homeodomain proteins to chromatin of the osteocalcin gene. Mol Cell Biol 24(20):9248–9261. doi:10.1128/MCB.24.20.9248-9261.2004

9. Chen D, Harris MA, Rossini G, Dunstan CR, Dallas SL, Feng JQ, Mundy GR, Harris SE (1997) Bone morphogenetic protein 2 (BMP-2) enhances BMP-3, BMP-4, and bone cell differentiation marker gene expression during the induction of mineralized bone matrix formation in cultures of fetal rat calvarial osteoblasts. Calcif Tissue Int 60(3):283–290

10. Hassan MQ, Tare RS, Lee SH, Mandeville M, Morasso MI, Javed A, van Wijnen AJ, Stein JL, Stein GS, Lian JB (2006) BMP2 commitment to the osteogenic lineage involves activation of Runx2 by DLX3 and a homeodomain transcriptional network. J Biol Chem 281(52):40515–40526. doi:10.1074/jbc.M604508200

11. Ulsamer A, Ortuno MJ, Ruiz S, Susperregui AR, Osses N, Rosa JL, Ventura F (2008) BMP-2 induces Osterix expression through up-regulation of Dlx5 and its phosphorylation by p38. J Biol Chem 283(7):3816–3826. doi:10.1074/jbc.M704724200

12. Cho TJ, Gerstenfeld LC, Einhorn TA (2002) Differential temporal expression of members of the transforming growth factor beta superfamily during murine fracture healing. J Bone Miner Res 17(3):513–520. doi:10.1359/jbmr.2002.17.3.513

13. Gurkan UA, Gargac J, Akkus O (2010) The sequential production profiles of growth factors and their relations to bone volume in ossifying bone marrow explants. Tissue Eng Part A 16(7):2295–2306. doi:10.1089/ten.TEA.2009.0565

14. Prince M, Banerjee C, Javed A, Green J, Lian JB, Stein GS, Bodine PV, Komm BS (2001) Expression and regulation of Runx2/Cbfa1 and osteoblast phenotypic markers during the growth and differentiation of human osteoblasts. J Cell Biochem 80(3):424–440

15. Shea CM, Edgar CM, Einhorn TA, Gerstenfeld LC (2003) BMP treatment of C3H10T1/2 mesenchymal stem cells induces both chondrogenesis and osteogenesis. J Cell Biochem 90(6):1112–1127. doi:10.1002/jcb.10734

16. Cho HY, Srinivasan A, Hong J, Hsu E, Liu S, Shrivats A, Kwak D, Bohaty AK, Paik HJ, Hollinger JO, Matyjaszewski K (2011) Synthesis of biocompatible PEG-Based star polymers with cationic and degradable core for siRNA delivery. Biomacromolecules 12(10):3478–3486. doi:10.1021/bm2006455

17. Averick SE, Paredes E, Irastorza A, Shrivats AR, Srinivasan A, Siegwart DJ, Magenau AJ, Cho HY, Hsu E, Averick AA, Kim J, Liu S, Hollinger JO, Das SR, Matyjaszewski K (2012) Preparation of cationic nanogels for nucleic acid delivery. Biomacromolecules 13(11):3445–3449. doi:10.1021/bm301166s

18. McNaughton BR, Cronican JJ, Thompson DB, Liu DR (2009) Mammalian cell

penetration, siRNA transfection, and DNA transfection by supercharged proteins. Proc Natl Acad Sci U S A 106(15):6111–6116. doi:10.1073/pnas.0807883106

19. Turchinovich A, Zoidl G, Dermietzel R (2010) Non-viral siRNA delivery into the mouse retina in vivo. BMC Ophthalmol 10:25. doi:10.1186/1471-2415-10-25

20. Komori T (2006) Regulation of osteoblast differentiation by transcription factors. J Cell Biochem 99(5):1233–1239. doi:10.1002/jcb.20958

21. Franceschi RT, Xiao G (2003) Regulation of the osteoblast-specific transcription factor, Runx2: responsiveness to multiple signal transduction pathways. J Cell Biochem 88 (3):446–454. doi:10.1002/jcb.10369

22. Schmittgen TD, Livak KJ (2008) Analyzing real-time PCR data by the comparative CT method. Nat Protoc 3(6):1101–1108. doi:10.1038/nprot.2008.73

23. Wang YH, Liu Y, Maye P, Rowe DW (2006) Examination of mineralized nodule formation in living osteoblastic cultures using fluorescent dyes. Biotechnol Prog 22(6):1697–1701. doi:10.1021/bp060274b

24. Stephens AS, Stephens SR, Morrison NA (2011) Internal control genes for quantitative RT-PCR expression analysis in mouse osteoblasts, osteoclasts and macrophages. BMC Res Notes 4:410. doi:10.1186/1756-0500-4-410

25. Vandesompele J, De Preter K, Pattyn F, Poppe B, Van Roy N, De Paepe A, Speleman F (2002) Accurate normalization of real-time quantitative RT-PCR data by geometric averaging of multiple internal control genes. Genome Biol 3(7):RESEARCH0034

Methods in Molecular Biology (2014) 1202: 161–171
DOI 10.1007/7651_2013_38
© Springer Science+Business Media New York 2013
Published online: 24 October 2013

Mimicking Bone Microenvironment for Directing Adipose Tissue-Derived Mesenchymal Stem Cells into Osteogenic Differentiation

ZuFu Lu, Seyed-Iman Roohani-Esfahani, and Hala Zreiqat

Abstract

Adipose tissue-derived mesenchymal stem cells (ASCs) have become an increasingly interested cell source for the scientists in the fields of stem cell biology and regenerative medicine. ASCs have already been used in a number of clinical trials, and some successful outcomes have been reported in bone tissue regeneration. Here we describe the protocols which mimic the factors in bone healing microenvironment, including inflammation burst, osteoblasts, and bone biomimetic scaffolds to direct ASCs into osteogenic differentiation.

Keywords: Adipose tissue-derived mesenchymal stem cells, Bone microenvironment, Osteogenic differentiation, TNF-α, Osteoblasts, Biomimetic scaffolds

1 Introduction

Adipose tissue-derived mesenchymal stem cells (ASCs) were first identified and characterized by Zuk and his colleagues (1), and since then they have become an increasingly interested cell source for the scientists in the fields of stem cell biology and regenerative medicine. ASCs have the multilineage differentiation capability such as adipogenic, osteogenic, and chondrogenic lineage under an appropriate condition, similar to those of bone marrow-derived mesenchymal stem cells (BM-MSCs) (2). However, the less invasive harvesting procedure of ASCs and their larger quantity of yield place them in a unique position relative to other MSCs (3). ASCs have already been used in a number of clinical trials, and some successful outcomes have been reported, especially in tissue reconstruction (4, 5).

Bone tissue has the innate capability for self-repair or regeneration throughout life; however, the healing process fails in complicated pathological fractures or large bone defects, where the bridging of nonunion and critical-sized bone defects remains to be a significant challenge for orthopedic surgeons around the world.

Various in vitro and in vivo bone tissue engineering approaches utilizing ASCs have been reported (3, 6), and the key principle for these approaches is to provide appropriate signals to instruct ASCs into osteogenic lineage. Various factors in tissue microenvironment play a vital role in bone tissue repair and regeneration upon injury, and mimicking such a microenvironment has been shown to be a feasible approach to control stem cell fate into the designated lineage (7–11). Bone regeneration involves three phases: namely, inflammatory phase (3-day burst); reparative phase; and remodelling phase. Factors within this microenvironment, such as inflammatory cytokines, growth factors, and cells (e.g., osteoblasts), closely participate in the bone healing or regeneration process (12, 13).

In this chapter, we describe the different strategies involved in directing ASCs into osteogenic differentiation, which mimic the factors that play a role in bone healing/repairing microenvironment including inflammatory burst, (pre-)osteoblasts, and trabecular bone and biomimetic scaffolds.

2 Materials

1. ASCs (Invitrogen, USA).

2. α-Minimal Essential Medium (α-MEM, L-glutamine (+), Gibco Laboratories, USA).

3. Fetal calf serum (FCS, Gibco Laboratories, USA).

4. Penicillin–Streptomycin (5,000 U/ml) (Gibco Laboratories, USA).

5. L-Ascorbic acid phosphate magnesium salt (Wako Pure Chemicals, Osaka, Japan).

6. β-Glycerophosphate (Sigma, USA).

7. Tumor necrosis factor-alpha (TNF-α, Sigma, USA).

8. Bone chips (normal human trabecular bone from the young patients undergone corrective surgery).

9. Transwell inserts (0.4 μm pore size, Millipore, USA).

10. Phosphate-buffered saline (PBS).

11. TrypLE™ Express (Sigma, USA).

12. 0.25 % Trypsin–EDTA (1×) (Gibco Laboratories, USA).

13. Ammonium phosphate dibasic ((NH_4)$_2$$HPO_4$, Sigma-Aldrich reagent grade, ≥98 %, USA) solution: Dissolve 75.27 g of ammonium phosphate dibasic in 600 ml deionized water (beaker B).

14. Calcium nitrate tetrahydrate ($Ca(NO_3)_2 \cdot 4H_2O$, Sigma-Aldrich, >99 %, USA) solution: Add 1,000 ml deionized

water to a beaker, and dissolve 226.70 g of calcium nitrate tetrahydrate (beaker A).

15. Ammonium hydroxide solution (NH_4OH, Sigma-Aldrich, USA).

16. Hydrochloric acid (HCl, Sigma-Aldrich, USA).

17. Sodium hydroxide (NaOH, Sigma-Aldrich, USA).

18. Polycaprolactone granules (PCL ($C_6H_{10}O_2$)$_n$, MW:70,000–90,000, Sigma-Aldrich, USA).

19. Chloroform solution ($CHCl_3$, Sigma-Aldrich, ≥ 99.5 %, USA).

20. Polyvinyl alcohol (PVA [$-CH_2CHOH-$]$_n$, Sigma-Aldrich, USA).

21. Ethanol and acetone (Sigma-Aldrich, USA).

22. Calibrate a pH meter at pH values of 4, 7, and 10 (UltraBasic, UB-10).

23. Hot plate and magnet stirrer.

24. Thermometer (max 100 °C).

25. Fume hood (AirScience, PureAir™, Australia).

26. Filter paper.

27. Alumina crucible.

28. Oven and furnace (Carbolite 1500, DFA 7000, Germany).

29. Planetary ball mill (Retsch 400 PM, Germany).

30. Stainless steel sieve (Retsch, −25 µm, Germany).

31. Fully reticulate polyurethane foam (The Foam Booth, Sydney, Australia).

32. Compressed air.

33. A programmable electric furnace (Carbolite 1500, Germany).

34. Sonicator (MTI instrument, Australia).

35. Fume hood (AirScience, PureAir™, Australia).

36. Vacuum oven (MTI corporation, DFA 7000).

3   Methods

3.1   Three Days of TNF-α Boost Induces Osteogenic Differentiation of ASCs

All solutions and equipment that come in contact with the cells must be sterile. Always use proper sterile technique, and work in a laminar flow hood. ASCs at passage 4 were used for osteogenic differentiation.

1. Culture medium consisting of complete α-MEM and osteogenic medium: Complete α-MEM medium is the basal α-MEM medium supplemented with 10 % (v/v) FCS and penicillin–streptomycin (50 U/ml), while the osteogenic

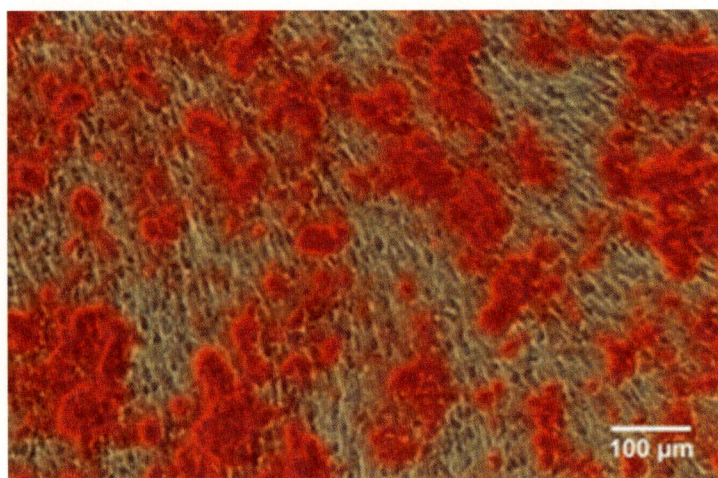

Fig. 1 TNF-α preconditioning induced osteogenic differentiation of ASCs. Three days of TNF-α preconditioning promotes the mineralization of ASCs at day 30 visualized by Alizarin red staining

medium is the complete α-MEM medium supplemented with 1 mM L-ascorbic acid phosphate magnesium salt and 10 mM β-glycerophosphate.

2. Quickly thaw one vial of ASCs by swirling it in a 37 °C water bath, and remove it once ice has completely melted.

3. When thawed, immediately transfer cells dropwise into a 75 cm² flask, which contains 15 ml of pre-warmed complete α-MEM medium.

4. Incubate at 37 °C, 5 % CO_2, and 90 % humidity, and allow cells to adhere for several hours (or overnight).

5. When the cells have attached to the growth surface, replace the medium with an equal volume of fresh, pre-warmed complete α-MEM medium.

6. Change the medium every 3–4 days till 80–90 % confluence.

7. ASCs at passage 4 are seeded on 12-well plate at a density of 8,000 cells/cm² and cultured in osteogenic medium with supplementation of TNF-α (1 ng/ml) for 3 days.

8. After 3 days, the medium is replaced with fresh osteogenic medium, and the osteogenic medium is refreshed every 3 days till designated time points for osteogenic differentiation assays. The mineralization of ASCs at day 30 was visualized by Alizarin red staining at day 30 (Fig. 1).

3.2 Human Primary Osteoblast-Like Cells Induce Osteogenic Differentiation of ASCs in an Indirect Co-culture System

All solutions and equipment that come in contact with the cells must be sterile. Always use proper sterile technique, and work in a laminar flow hood. Prepare complete α-MEM medium: Complete α-MEM medium is the basal medium α-MEM supplemented with 10 % (v/v) FCS, penicillin–streptomycin (50 U/ml), and 1 mM L-ascorbic acid phosphate magnesium salt.

3.2.1 HOB Isolation

1. Human trabecular bone chips are chopped into 1 mm^3 pieces and washed several times in PBS.

2. Digestion for 90 min at 37 °C with 0.02 % (w/v) trypsin in PBS.

3. Digested cells were cultured in complete α-MEM medium.

4. The cells are cultured at 37 °C with 5 % CO2, the medium is refreshed every 3 days until confluence, and then cells are passaged till passage 2 (see **Note 1**).

3.2.2 HOB–ASC Co-culturing

Cell culture inserts with a 0.4-μm pore-size inserts are used for indirect co-culture of human primary osteoblast-like cells (HOBs) and ASCs (Fig. 2).

1. HOBs are seeded on the 6-well culture plates with a cell density of 10,000 cells/cm^2.

2. ASCs at passage 4 are freshly digested from the culture flask and seeded on the cell culture inserts with a cell density of 8,000 cells/cm^2 (~50 % confluence).

3. The inserts are placed in 6-well plates for the indirect co-culture of ASCs with HOBs.

4. The medium is replaced with fresh complete α-MEM medium every 3 days till designated time points for osteogenic differentiation assays.

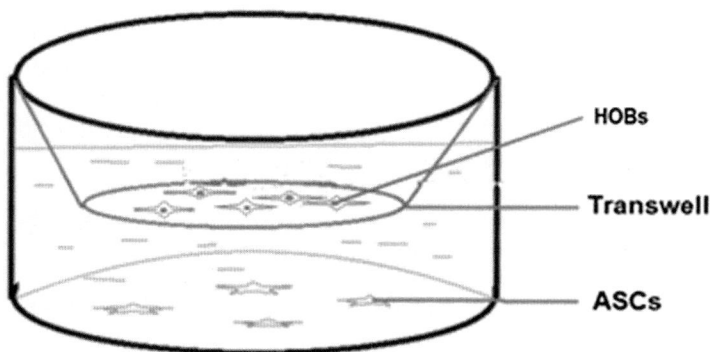

Fig. 2 Illustration of indirect ASC–HOB co-culture system. HOBs promote early osteogenic differentiation of ASCs in an indirect co-culture system

3.3 Trabecular Bone Structure Mimetic (14) Scaffold Mediates Osteogenic Differentiation of ASCs

3.3.1 Preparation of Rod-Shaped Hydroxyapatite Nanoparticles

1. Place beaker A on a stirrer/hot plate.

2. Place a thermometer in beaker A, and adjust the temperature to 80 ± 2 °C. (It is important to keep the temperature at 80 °C during the entire process (see **Note 2**).)

3. Cover beaker A with a plastic wrap (see **Note 3**).

4. Place beaker B on a stirrer/hot plate, and adjust the temperature to 90 ± 2 °C (see **Note 4**).

5. Transfer beakers A and B in fume hood, and put the pH meter probe inside the beaker A. pH should be in the range of 5–7.

6. Adjust the pH to 6.5 by adding 2 M NH_4OH solution and 1 M HCl.

7. By using a pipette, start adding solution from beaker B to beaker A in a dropwise manner (see **Note 5**).

8. Once solution in beaker B is used completely, seal beaker A with a plastic wrap and leave it for 1-h mixing (see **Note 6**).

9. Remove the beaker from stirrer, and maintain at room temperature overnight to initiate the aging process (see **Note 7**).

10. After the aging process, filter and wash the white precipitates by using filter paper and distilled water and transfer the gel into a clean glass beaker (see **Note 8**).

11. Transfer the beaker to an electric oven, and set the temperature to 90 °C.

12. After drying for 2 days, transfer the dried agglomerates to an alumina crucible.

13. Put the alumina crucible in an electric furnace and calcine agglomerates at 700 °C for 2 h.

14. Grind the agglomerates with mortar and pestle. Transfer the de-agglomerated particles to a beaker containing 100 % ethanol, and sonicate them for 30 min. The obtained powder morphology is shown in Fig. 3 (see **Note 9**).

3.3.2 Preparation of HA/β-TCP (40:60 wt%) Scaffolds

1. Place beaker A on a stirrer/hot plate.

2. Place a thermometer in beaker A, and adjust the temperature to 80 ± 5 °C. (It is very important to keep the temperature around 80 °C throughout the entire process (see **Note 2**).)

3. Cover beaker A with a plastic wrap.

4. Place beaker B on a stirrer/hot plate, and adjust the temperature to 90 ± 2 °C (see **Note 4**).

5. Transfer beakers A and B into the fume hood, and place the pH meter probe inside beaker A. pH should be in the range of 5 and 7.

Fig. 3 SEM images of rod-shaped HA nanoparticles

6. Adjust pH to 8.5 by adding 2 M NH$_4$OH solution and 1 M HCl.

7. Using a pipette, start adding the solution from beaker B to beaker A in a dropwise manner.

8. Adding solution from beaker B to beaker A leads to the formation of white precipitates and the drastic decrease in pH to 3. Therefore it is critical to perform this step in a drop-by-drop manner while maintaining the pH at 8.5 (see **Note 5**).

9. Once solution in beaker B is used, seal beaker A with a plastic wrap and leave for 1 h to mix (see **Note 6**).

10. Remove the beaker from stirrer, and place at room temperature overnight to initiate the aging process (see **Note 7**).

11. After the aging process, filter and wash the white precipitates by using the filter paper and distilled water and transfer the resultant gel into a clean glass beaker (see **Note 8**).

12. Transfer the beaker to an electric oven, and set the temperature to 90 °C.

13. After drying for 2 days, transfer the dried agglomerates to an alumina crucible.

14. Place the alumina crucible in an electric furnace and calcine agglomerates at 900 °C for 2 h.

15. Crush the large agglomerates by mortar and pestle to achieve a finer agglomerate, and then transfer to a zirconia jar. Add ten zirconia balls ($d = 2.5$ cm), 50 g of fine agglomerates, and 150 ml ethanol in each jar. Mount the jars on the planetary ball mill, and grind them at 150 rpm for 2 h. Open the jars, and pour the slurry on a clean beaker.

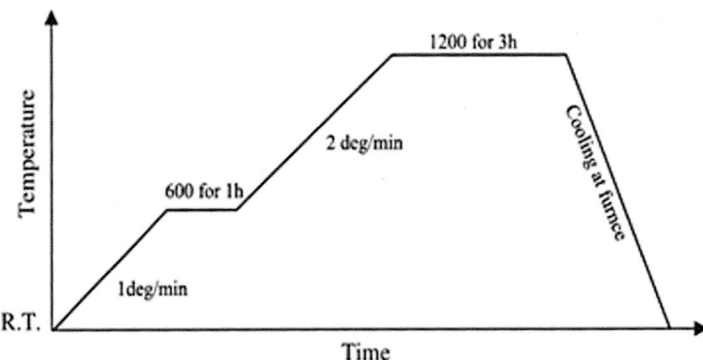

Fig. 4 Heat treatment program designed for burning out the polyurethane templates and sintering the HA/β-TCP (hydroxyapatite/beta-tricalcium phosphate) scaffolds

16. Transfer the beaker to an oven, and set the temperature to 90 °C.

17. Sieve the dried powder with a stainless steel sieve (−75 μm).

18. Add the powder to PVA solution to make a 30 wt% suspension (see **Note 10**).

19. Clean the foams, and immerse the foam into 1 M NaOH solution for 30 min.

20. Take out the foams from the NaOH solution, and dry them in oven at 37 °C for 5 h.

21. Immerse foam into the powder/PVA suspension, and compress it slightly to facilitate slurry penetration. Remove the foam, and squeeze out the excess slurry and subsequently blasts with compressed air to ensure a uniform coating of slurry (see **Note 11**).

22. Transfer the coated foam to an oven, and dry them at 37 °C for 24 h.

23. After drying, place the coated foams on an alumina tray, put the tray into a programmable electric furnace, and fire the foams by using four-stage schedule (Fig. 4).

3.3.3 Nanocomposite Coating Process on HA/β-TCP Scaffolds

1. Dissolve PCL granules in chloroform solution at a concentration of 7 % (w/v) by stirring PCL and chloroform in a glass beaker using a magnet stirrer (see **Note 12**).

2. Disperse 3 vol.% of rod-shaped HA nanoparticles in 10 ml chloroform by sonication for 15 min, and then add the suspension to the dissolved PCL.

3. Clean the HA/β-TCP scaffolds with ethanol and acetone, and dry them in oven at 37 °C.

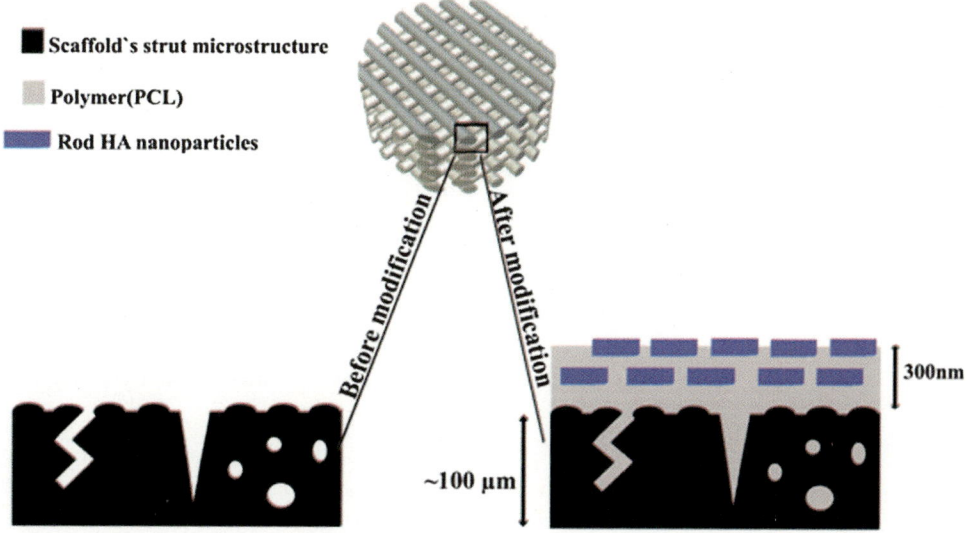

Fig. 5 Schematic diagram for coating a nanocomposite layer over the ceramic struts

4. After 24-h homogenization of nanoparticles and PCL suspension, immerse the HA/β-TCP scaffolds into suspension for 1 min, then take out the scaffolds, and remove the extra suspension using compressed air (Fig. 5) (see **Note 13**).

5. Dry the coated scaffolds in an oven at 37 °C for 3 days followed by 1 day in vacuum oven for 5 h.

*3.3.4 ASCs Seeded on Scaffolds for Osteogenic Differentiation (See **Note 14**)*

1. Before cell seeding, all the scaffolds are decontaminated by soaking them twice in 70 % ethanol for 30 min each, followed by rinsing three times with PBS for 5 min each before exposing to ultraviolet light for 30 min.

2. Prepare osteogenic medium: The basal medium α-MEM is supplemented with 10 % (v/v) FCS, penicillin–streptomycin (50 U/ml), 1 mM L-ascorbic acid phosphate magnesium salt, and 10 mM β-glycerophosphate.

3. A suspension of 100 μl culture medium containing 150,000 ASCs (at passage 4) was dropped gently onto the scaffolds and incubated for 90 min at 37°C to allow cells to attach to the scaffolds, before flooding with the cell culture medium.

4. The medium is replaced with fresh osteogenic medium and replenished every 3 days till designated time points for osteogenic differentiation assays.

4 Notes

1. All primary HOBs used in the experiments are at passage 2. As the HOBs at later passage may gradually lose their osteogenic phenotype, their capability of inducing ASCs into osteogenic lineage might be compromised.

2. Instead of addition of calcium nitrate to distilled water at 80 °C, dissolve the calcium nitrate at room temperature first, and then adjust the solution's temperature to 80 °C. This will reduce the temperature fluctuations during addition of calcium nitrate.

3. Cover the beaker by a cling wrap, and seal it by using a rubber band.

4. After dissolving the ammonium phosphate dibasic in distilled water, as an alternative way, you may transfer the beaker to microwave chamber, heat up the solution just before formation of bubbles, and adjust the temperature to 90 °C. Always keep the beaker covered by cling wrap to prevent evaporation of the solution.

5. Adding solution from beaker B to beaker A causes the formation of white precipitates and a drastic decrease in pH. Therefore it is critical to add solution from beaker B to beaker A in a dropwise manner to keep pH and temperature of the solution A at 6.5 and 80 °C, respectively. This process should usually take 3–4 h to finish.

6. During mixing make sure that the heater is off till the solution cools off to room temperature.

7. Make sure that the beaker is completely sealed during the aging process. Evaporation of mother solution can affect the aging process and results in a powder with defective crystal structure.

8. Filtering can also be done by using a vacuum pump. After filtration, wash the gels with distilled water at least three times.

9. The aim of grinding is to refine the agglomerates not refining the powder; therefore the process should be done as gently as possible to reduce the chance of powder refining and hence changing the powder morphology. After sonication for 30 min, a part of powder is deposited at the bottom of the beaker and other part is still floating. Using a plastic pipette collect the solution containing floated particles, pour it on a glass slide, and transfer the glass slide to the oven. Repeat the sonication and grinding processes till all of the deposited particles become floatable.

10. The homogeneity of PVA solution and powder is crucial to get a uniform coating. Mixing time at least should be 4 h to achieve an acceptable consistency.

11. Make sure that all the excess slurry is out of the foam. The remaining excess slurry in the polymer foam can compromise the interconnectivity between the pores and total porosity. One way to detect the remained slurry in the foam is to weigh the coated foams.

12. The chloroform vapor is very toxic and can be extremely harmful. Read the MSDS instruction for safety practices and its related hazards.

13. Chloroform can dissolve the plastic-based gloves, so make sure not to touch the coated scaffolds with gloves in hand.

14. It is important to consider combining different factors to induce the full and efficient osteogenic differentiation of ASCs. The synergistic effects of bone biomimetic scaffolds and osteoblasts on osteogenic differentiation of ASCs are observed, and one single factor might only be able to induce early osteogenic differentiation of ASC (10).

References

1. Zuk PA, Zhu M, Ashjian P et al (2002) Human adipose tissue is a source of multipotent stem cells. Mol Biol Cell 13:4279–4295

2. Seda Tigli R, Ghosh S, Laha MM et al (2009) Comparative chondrogenesis of human cell sources in 3D scaffolds. J Tissue Eng Regen Med 3:348–360

3. Levi B, Longaker MT (2011) Concise review: adipose-derived stromal cells for skeletal regenerative medicine. Stem Cells 29:576–582

4. Mauney JR, Nguyen T, Gillen K et al (2007) Engineering adipose-like tissue in vitro and in vivo utilizing human bone marrow and adipose-derived mesenchymal stem cells with silk fibroin 3D scaffolds. Biomaterials 28:5280–5290

5. Wilson A, Butler PE, Seifalian AM (2011) Adipose-derived stem cells for clinical applications: a review. Cell Prolif 44:86–98

6. Szpalski C, Barbaro M, Sagebin F et al (2012) Bone tissue engineering: current strategies and techniques–part II: cell types. Tissue Eng Part B Rev 18:258–269

7. Giannoni P, Lazzarini E, Ceseracciu L et al (2012) Design and characterization of a tissue-engineered bilayer scaffold for osteochondral tissue repair. J Tissue Eng Regen Med. doi:10.1002/term.1651

8. Hwang NS, Varghese S, Lee HJ et al (2013) Biomaterials directed in vivo osteogenic differentiation of mesenchymal cells derived from human embryonic stem cells. Tissue Eng Part A 16(15–16):1723–1732. doi:10.1089/ten.tea.2013.0064

9. Lu Z, Roohani-Esfahani SI, Kwok PC et al (2011) Osteoblasts on rod shaped hydroxyapatite nanoparticles incorporated PCL film provide an optimal osteogenic niche for stem cell differentiation. Tissue Eng Part A 17:1651–1661

10. Lu Z, Roohani-Esfahani SI, Wang G et al (2012) Bone biomimetic microenvironment induces osteogenic differentiation of adipose tissue-derived mesenchymal stem cells. Nanomedicine 8:507–515

11. Lu Z, Wang G, Dunstan CR et al (2013) Activation and promotion of adipose stem cells by tumour necrosis factor-alpha preconditioning for bone regeneration. J Cell Physiol 228:1737–1744

12. Almeida CR, Vasconcelos DP, Goncalves RM et al (2012) Enhanced mesenchymal stromal cell recruitment via natural killer cells by incorporation of inflammatory signals in biomaterials. J R Soc Interface 9:261–271

13. Mountziaris PM, Mikos AG (2008) Modulation of the inflammatory response for enhanced bone tissue regeneration. Tissue Eng Part B Rev 14:179–186

14. Bae SE, Bhang SH, Kim BS et al (2012) Self-assembled extracellular macromolecular matrices and their different osteogenic potential with pre-osteoblasts and rat bone marrow mesenchymal stromal cells. Biomacromolecules 13:2811–2820

Methods in Molecular Biology (2014) 1202: 173–184
DOI 10.1007/7651_2013_52
© Springer Science+Business Media New York 2013
Published online: 27 November 2013

Cultivation of Human Bone-Like Tissue from Pluripotent Stem Cell-Derived Osteogenic Progenitors in Perfusion Bioreactors

Giuseppe Maria de Peppo, Gordana Vunjak-Novakovic, and Darja Marolt

Abstract

Human pluripotent stem cells represent an unlimited source of skeletal tissue progenitors for studies of bone biology, pathogenesis, and the development of new approaches for bone reconstruction and therapies. In order to construct in vitro models of bone tissue development and to grow functional, clinical-size bone substitutes for transplantation, cell cultivation in three-dimensional environments composed of porous osteoconductive scaffolds and dynamic culture systems—bioreactors—has been studied. Here, we describe a stepwise procedure for the induction of human embryonic and induced pluripotent stem cells (collectively termed PSCs) into mesenchymal-like progenitors, and their subsequent cultivation on decellularized bovine bone scaffolds in perfusion bioreactors, to support the development of viable, stable bone-like tissue in defined geometries.

Keywords: Human embryonic stem cells, Human induced pluripotent stem cells, Mesenchymal-like progenitors, Osteogenic differentiation, Osteogenic medium, Bone scaffolds, Cell seeding, Perfusion bioreactor, Medium flow rate, Bone tissue development

1 Introduction

Bioengineered human tissues have a broad range of applications in regenerative medicine, including reconstructive therapies, studies of development, disease modeling, and drug development and screening (1, 2). A variety of human stem cells from embryonic, fetal, and adult tissues are under investigation for the bioengineering of functional bone tissue substitutes (3). Among these, human pluripotent stem cells (hPSCs) with their unlimited growth and the potential to differentiate into any cell type of the human body, and immunocompatibility represent an unprecedented resource (4, 5). However, pluripotency also represents a challenge for reproducible, stable, and efficient induction of cells into the desired lineages.

For most mesodermal lineages, stepwise differentiation protocols were developed to reproduce the signals encountered by the pluripotent cells during early embryonic patterning, and guide the cell development in vitro (6). Similarly, osteogenic- and mesenchymal stem cell-like progenitors (MPs) were derived from hPSCs by a variety of protocols, involving two- or three-dimensional (3D) embryoid body cell cultures in the presence of inductive signals such as serum, growth factors, osteogenic factors, and coculture with primary stromal/osteogenic cells (reviewed in (1)). In these studies, osteogenic marker expression and limited (<1 mm) bone-like tissue formation were demonstrated either in vitro or in vivo. In some cases, cells formed teratomas after extended in vivo transplantation, suggesting the need for further optimization of the hPSC induction/tissue engineering protocols (7, 8).

Bone tissue engineering has been extensively studied with adult mesenchymal stem cells (MSCs) (reviewed in (3, 9)). The influences of various tissue engineering and in vitro culture parameters on bone development have been elucidated, such as delivery of osteoinductive factors, the properties of osteoconductive scaffolds (composition, structure, mechanical properties), and the culture in static or dynamic environments, with their specific mass transport and gas exchange properties, and the possibilities to deliver select biochemical and biophysical stimuli (3).

Our group has developed a biomimetic tissue-engineering strategy for the cultivation of functional, anatomically shaped bone substitutes, by culturing MSCs from the bone marrow on 3D porous scaffolds resembling the matrix of native bone in bioreactors providing interstitial flow of culture medium (10). For smaller (<1 cm) cylindrical bone constructs, we have defined the influences of initial cell seeding density, fluid perfusion rate, and bone scaffold architecture on in vitro bone development (11–13).

Herein, we describe this in vitro bone development model adjusted for growing bone-like tissue from hPSCs (14, 15). Based on previous studies (16, 17), we hypothesized that hPSCs-derived MPs will form bone-like tissue under conditions optimized with MSCs (11, 12), and that the osteogenic induction of MPs should thus be followed by the conditions supporting bone formation. Indeed, we found that the stepwise induction of MPs, and their subsequent seeding and culture on decellularized bovine bone scaffolds in osteogenic medium under constant perfusion resulted in the formation of viable, stable bone-like tissue of defined geometries (14, 15).

2 Materials

Dulbecco's modified eagle medium (DMEM), KnockOut DMEM (KO-DMEM), KnockOut Serum Replacement (KO-SR), Dulbecco's phosphate buffered saline solution (DPBS), GlutaMAX solution (100×), nonessential amino acids (100×), beta-mercaptoethanol, penicillin–streptomycin solution (100×, 100 U/ml), trypsin/

ethylenediaminetetraacetic acid (EDTA; 0.25 %), bovine serum albumin fraction V (BSA), fetal bovine serum (FBS), insulin, sodium pyruvate (100×), and basic fibroblast growth factor (bFGF) were purchased from Life Technologies. HyClone FBS, Tris buffer, proteinase K and phosphate buffered saline (PBS, 10×) were purchased from Fisher Scientific. Antibodies for flow cytometry were from BD Biosciences. Bovine bone joints were from Green Village Packing. All other chemicals were purchased from Sigma-Aldrich unless otherwise noted.

Prepare all solutions using tissue-culture grade water, unless otherwise instructed. Prepare culture media fresh for weekly media changes, and differentiation media fresh prior to each medium change. Culture media are prepared complete with the addition of serum/serum replacement, antimicrobials, cytokines, growth factors, and other supplements, filter-sterilized and stored at 4–8 °C. Culture and differentiation media are warmed to 37 °C prior to media changes and scaffold and bioreactor conditioning.

2.1 hPSC Cultivation and Induction into Mesenchymal-Progenitors

1. hPSC culture medium:
   Prepare medium by combining 80 % KO-DMEM with 20 % KO-SR (vol/vol), and adding 10 ng/ml bFGF, 2 mM Gluta-MAX, 0.1 mM nonessential amino acids, 0.1 mM beta-mercaptoethanol, and 100 U/ml (1×) penicillin–streptomycin (*see* **Note 1**).

2. Gelatin-coated plates:
   Dissolve gelatin 0.1 % (wt/vol) in tissue culture water, filter-sterilize, and store at room temperature. Pipet 2 ml of 0.1 % gelatin solution per well of 6-well tissue culture plates (Nunclon) or 0.5 ml of gelatin solution per 4-well/24-well plates and incubate at least 2 h at room temperature. Gelatin-coated tissue culture plates can be stored for several weeks in the incubator at 37 °C.

3. Mouse embryonic fibroblasts (MEF) medium:
   Make-up MEF medium by combining 90 % DMEM with 10 % FBS (vol/vol), and adding 100 U/ml penicillin–streptomycin.

4. MEF feeder layers:
   Thaw one vial of MEF (irradiated, strain CF1, GlobalStem) and transfer into a tube with 5–10 ml of MEF medium at 37 °C. Centrifuge the cell suspension at $250 \times g$ for 5 min, resuspend in MEF medium and seed at 150,000–300,000 cells/well of gelatin-coated 6 well plate or 40,000–75,000 cells/well of gelatin-coated 4 well plate. MEF feeder layers are used after 1–3 days of culture (at 37 °C in a humidified atmosphere containing 5 % CO_2—standard conditions) for seeding undifferentiated hPSC cells (*see* **Note 2**).

5. Mesenchymal induction medium:
Prepare induction medium by combining 80 % KO-DMEM with 20 % HyClone FBS (vol/vol), and adding 2 mM Gluta-MAX, 0.1 mM nonessential amino acids, 0.1 mM beta-mercaptoethanol, and 100 U/ml penicillin–streptomycin.

2.2 Characterization of hPSC-Mesenchymal Progenitor Surface Antigen Expression and Osteogenic Potential

1. Flow cytometry buffer and antibodies:
Prepare flow cytometry buffer by combining DPBS with 0.5 % BSA (vol/vol), 100 U/ml penicillin–streptomycin, 2 mM EDTA, and 20 mM glucose, sterile filter and store at 4–8 °C. Use a combination of antibodies to assess the mesenchymal-like surface antigen profile of progenitors derived from hPSC (14), for example: Tra1-60 PE (catalog no. 560193), SSEA1 PE (560142), SSEA4 V450 (561156), CD14 PE (561707), CD31 PE (555446), CD34 PE (555822), CD44 PE (561858), CD45 PE (560975), CD73 FITC (561254), CD90 PE (555596), and appropriate isotype controls (BD Biosciences). Dilute each antibody to a final concentration of 2 μl per 100 μl of flow cytometry buffer prior to staining. Alternatively, prepare several antibody dilutions to determine the lowest antibody concentration resulting in strong fluorescent signal for the staining.

2. Osteogenic medium:
Prepare osteogenic medium by combining 90 % DMEM with 10 % HyClone FBS (vol/vol), and adding 100 U/ml penicillin–streptomycin, 1 μM dexamethasone, 10 mM beta-glycerophosphate, and 50 μM ascorbic acid-2-phosphate.

3. Control medium:
Prepare control medium by combining 90 % DMEM with 10 % HyClone FBS (vol/vol), and adding 100 U/ml penicillin–streptomycin.

4. Alkaline phosphatase staining components:
The components of Fast blue RR Salt staining kit (Sigma-Aldrich) are prepared according to the manufacturer's instructions.

Prepare citrate working solution by diluting 2 ml citrate concentrate solution with 98 ml with deionized H_2O.

Prepare acetone fixative solution (citrate buffered acetone, 60 %) by adding 2 volumes of room temperature citrate working solution to 3 volumes of acetone under constant stirring (*see* **Note 3**).

Dissolve one capsule of Fast blue RR Salt in 48 ml of room temperature H_2O, and add 2 ml of Naphthol AS-MX Phosphate Alkaline Solution (*see* **Note 4**).

Prepare fresh filtered Mayor's Hematoxylin Solution.

5. Von Kossa staining components:
Prepare 3.7 % formaldehyde in PBS (vol/vol) and store at room temperature.

Dissolve AgNO$_3$ in distilled H$_2$O to make a 2 % (w/vol) staining solution (*see* **Note 5**).

2.3 Decellularized Bovine Bone Scaffolds

1. Bovine trabecular bone:
Obtain metacarpal joints of 2-week to 4-month-old calves from local butcher. Remove the soft tissues from the joint using scalpel and drill into the subchondral trabecular bone region to obtain cylinders of appropriate diameter (4–8 mm) (*see* **Note 6**).

2. 0.1 % EDTA solution in PBS:
Dissolve 1 g of EDTA in 1,000 ml PBS and store at room temperature.

3. 0.1 % EDTA solution in Tris:
Dissolve 1 g of EDTA in 1,000 ml of 10 mM Tris buffer and store at room temperature.

4. 0.5 % SDS solution in Tris:
Dissolve 0.5 ml of SDS in 1,000 ml of 10 mM Tris buffer and store at room temperature.

5. DNAse/RNase solution in Tris:
Dissolve one vial (2,000 Kunitz, ≥0.5 mg total protein) of freeze-dried DNAse (Sigma Aldrich) and one piece (0.1–0.15 mg) of dust RNAse A (Sigma Aldrich) per each 40 ml of 10 mM Tris in sterile H$_2$O.

2.4 Cultivation in Perfusion Bioreactors

1. Perfusion bioreactor components and tubing:
Gather the components of specific perfusion bioreactor system selected for the study. For instance, the components of custom-made round bioreactors used in our studies (14, 15) and the assembled bioreactor system are presented in Fig. 1 (11). For reusable bioreactor systems, disinfect the culture chambers/cassettes at the end of each experiment in 10 % hydrogen peroxide in water (vol/vol) for 30 min, wash thoroughly under stream of water, rinse with deionized water, and air-dry. Prior to the experiment, loosely assemble the bioreactors, connect new (medical-grade) tubing of appropriate length, fit the connectors for the stopcocks/syringes and sterilize by autoclaving.

2. Tools for bioreactor assembly and peristaltic pump components:
Prepare screws, hex-L-key, forceps, scissors, scalpels and other stainless-steel tools, and sterilize by autoclaving.
Assemble the peristaltic pump and transfer to the top/inside the incubator (depending on the selected bioreactor/pump system).

3. Osteogenic medium (prepared as in Section 2.2, item 2).

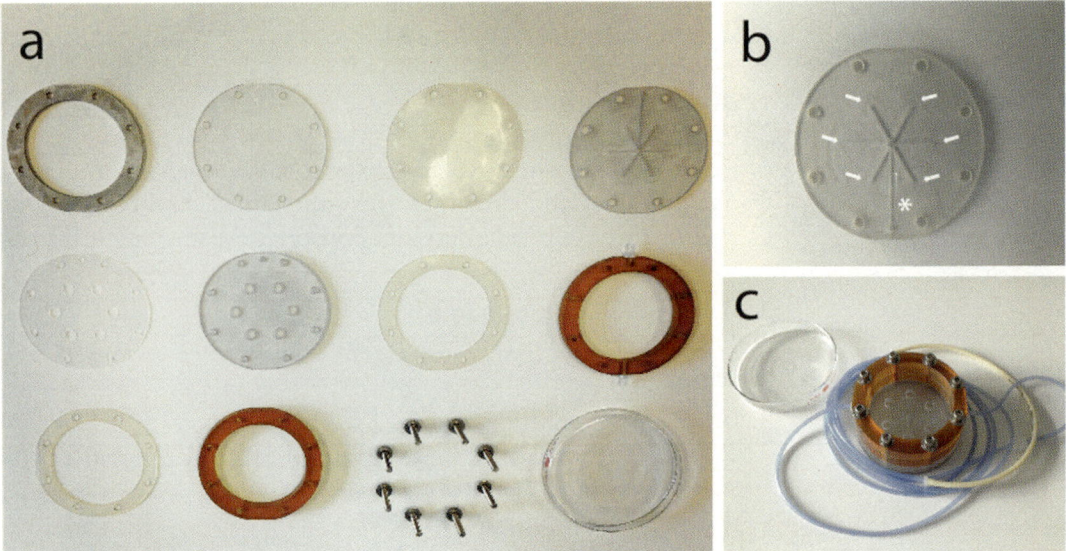

Fig. 1 Round bioreactor system for simultaneous perfusion of six cell-scaffold constructs. Components of the reusable round bioreactor system are cleaned, loosely assembled, and autoclaved prior to each use (**a**). The system holds six wells into which cell-scaffold constructs are tightly fitted. The culture medium flows to the central channel (*asterisk*) and splits six ways (*arrows*) to assure equal perfusion of the constructs (**b**). The bioreactor is covered with a glass cover and connected to a single perfusion loop (**c**) (11, 14, 15)

3   Methods

3.1   Culture and Induction of Pluripotent Stem Cells into Mesenchymal Progenitors

1. Prepare gelatin-coated plates, MEF medium, hPSC medium, and MEF feeder layers. Thaw hPSCs on fresh feeder layers (typically 1 frozen well of cells per 1 well of the culture plate) and culture in hPSC medium in the incubator at standard conditions. Manually remove the differentiated areas and split the colonies 1:1 to 1:3 for the first passages (depending on the growth of particular hPSC line) using standard techniques (mechanical splitting or enzymatic treatment). Expand the hPSCs to a fully confluent 6-well plate of colonies in hPSC medium (Fig. 2a).

2. Change the hPSC medium in confluent cultures to mesenchymal induction medium and incubate for 1 week in the incubator at standard conditions (Fig. 2b). Change media on days 3 and 6.

3. After 7 days, aspirate the culture medium, wash the cultures with DPBS, and detach the cells by incubating in trypsin/EDTA for 5 min at 37 °C. Count and seed the cells at high density (100,000 cells/cm^2) in gelatin-coated plates. Incubate the cultures in mesenchymal induction medium in standard conditions and change media twice a week.

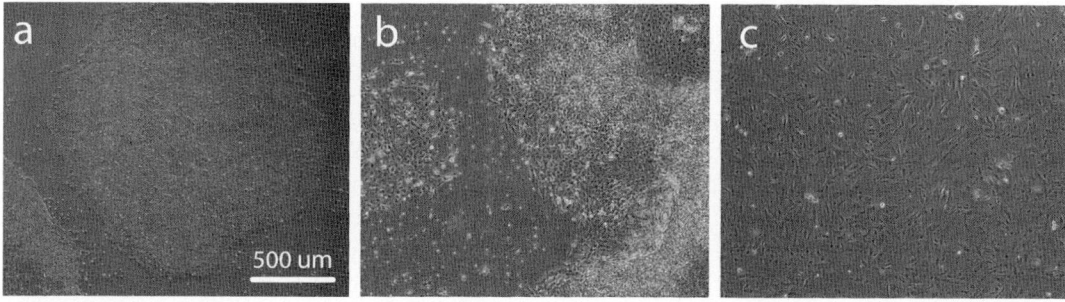

Fig. 2 Induction of hPSCs into MPs. Typical morphologies of undifferentiated hPSC cultures (**a**), overgrown hPSC cultures after 1 week in mesenchymal induction medium (**b**) and MPs after passaging in mesenchymal medium (**c**)

4. Upon reaching confluence, passage cells using trypsin/EDTA in mesenchymal induction medium at a density of $10,000/cm^2$ until they become homogenous for a fibroblastic-like morphology (passages 3–5 after induction, Fig. 2c) (*see* **Note 7**).

5. To assess the genomic stability of MPs following in vitro expansion, perform karyotpye analysis (for instance using a commercial provider, such as Cell Line Genetics).

3.2 Characterization of hPSC-Mesenchymal Progenitor Surface Antigen Expression

1. Detach expanded MPs with trypsin/EDTA, wash with mesenchymal induction medium and filter through a 70 mm cell strainer (BD Biosciences) to obtain a single-cell suspension.

2. Count the cells and resuspend 5×10^6 cells in 3 ml of a sterile flow cytometry buffer and transfer on ice.

3. Prepare 50 µl aliquots of diluted primary antibodies and dispense them into black 96-well plates (Corning), followed by the addition of 50 µl aliquots containing 8×10^4 cells for a total of 100 µl per well.

4. Incubate cells on ice in the dark for 30 min, wash in staining buffer, and analyze immediately on flow cytometer/sorter (BD Biosciences ARIA-IIu SOU Cell Sorter) configured with a 100 µm ceramic nozzle and operating at 20-psi sheath fluid pressure.

5. Collect and analyze data using the software provided by the system (BD Biosciences Diva 6.0 software), using a combination of standard procedures with fluorescence minus one and isotype controls.

3.3 Characterization of hPSC-Mesenchymal Progenitor Osteogenic Potential

1. Seed MPs at passages 3–5 (or any other selected passage) at a density of $10,000/cm^2$ in 24-well gelatin-coated plates in osteogenic and control media. Culture in standard conditions and change media twice a week.

2. At 2 and 4 weeks after seeding, evaluate alkaline phosphatase activity and matrix mineralization by von Kossa staining.

3. For alkaline phosphatase activity, rinse sample wells in PBS and then fix by immersing in citrate buffered acetone for 30 s. Rinse gently in deionized H_2O for 45 s. Do not allow the wells to dry. Then add alkaline-dye mixture and incubate at room temperature for 30 min in the dark. Rinse thoroughly with deionized H_2O and counterstain with Mayer's Hematoxylin solution for 10 min. Rinse abundantly with deionized H_2O and evaluate microscopically (*see* **Note 8**).

4. For assessing matrix mineralization by von Kossa staining, rinse sample wells in PBS and then fix in 3.7 % formaldehyde in PBS (vol/vol) for 30 min. Rinse twice with deionized H_2O, add 2 % $AgNO_3$ solution to cover the well surface and keep in the dark for 10 min. Rinse with deionized H_2O and then expose the samples to bright light (100 W) for 15–30 min. Then rinse again with deionized H_2O and either dehydrate quickly adding 95–100 % ethanol or store the samples in PBS at 4–8 °C (*see* **Note 9**).

5. Observe and document the positive blue-purple staining of alkaline phosphatase and black staining of accumulated minerals in osteogenic cultures by light microscopy (*see* **Note 10**).

3.4 Preparation of Decellularized Bovine Bone Scaffolds

1. Wash the bone cylinders under high-pressure water stream to remove the bone marrow (*see* **Note 11**).

2. Place bone cylinders into 50 ml centrifuge tubes to fill approximately one half of the tube volume. Wash the bone cylinders with 0.1 % solution of EDTA in PBS for 1 h, then with 0.1 % EDTA solution in 10 mM Tris for 12 h at 10–15 °C under agitation, and finally with 0.5 % SDS solution in 10 mM Tris for 18–24 h at room temperature under agitation. Rinse the samples in PBS 20–30 times until all bubbles disappear under agitation (15 min each wash).

3. Subsequently, incubate the bone cylinders in a solution of DNase and RNase in 10 mM Tris for 6 h at room temperature.

4. Rinse the decellularized bone cylinders two times with PBS, freeze-dry and cut to 4–5 mm thickness. Polish the bone disks to obtain scaffolds of 4 × 4 mm in diameter and thickness (*see* **Note 12**).

5. Weigh and measure each individual scaffold using a caliper to calculate the scaffold density. Select scaffolds in the range of 0.37–0.45 mg/mm^3 for bone engineering.

6. Sterilize scaffolds in 70 % ethanol (vol/vol) overnight at room temperature.

7. Condition the scaffolds in osteogenic medium overnight in the incubator before cell seeding.

3.5 Cell Seeding of Decellularized Bovine Bone Scaffolds

1. Expand MPs expressing mesenchymal surface markers and exhibiting osteogenic differentiation potential on gelatin-coated plates in mesenchymal medium (*see* **Note 13**).

2. Detach expanded MPs with trypsin/EDTA, wash with mesenchymal induction medium, determine the total cell number, centrifuge at $250 \times g$ for 5 min, and resuspend the cells to a seeding density of 30×10^6 cells/ml in osteogenic medium (*see* **Note 14**).

3. Place sterilized, conditioned scaffolds onto sterile gauze to blot the culture medium, and quickly transfer each scaffold per well of low attachment 6-well plates (Corning).

4. Pipet a 40 µl aliquot of the cell suspension (a total of 1.5×10^6 cells) onto each 4×4 mm bone scaffold carefully, allowing the cell suspension to penetrate the scaffold pores.

5. To facilitate uniform cell distribution, flip the scaffolds every 15 min for 1 h, and each time add 5 µl of the osteogenic medium to prevent the cells from drying out.

6. After 1 h, add 6 ml of osteogenic medium to each well and transfer the seeded scaffolds to the incubator. Incubate without disturbing in standard conditions for 3 days.

3.6 Cultivation of Cell-Seeded Scaffolds in Perfusion Bioreactors

1. On days 1–2 after cell seeding, prepare the bioreactor chambers. Tighten the assembled, sterilized perfusion bioreactors with sterilized tools working in the aseptic conditions (laminar flow hood) and fit the three-way stopcocks with syringes into the perfusion loop connectors. Fill the chambers with medium and place them in the incubator for overnight conditioning. Attach to the peristaltic pump and start the medium flow.

2. On day 3 after seeding, harvest some of the seeded constructs for the evaluation of cell viability, cell seeding efficiency, cell distribution, and other analyses at the start of perfusion culture. Transfer some of the constructs in new low attachment 6-well plates for static culture controls, and add to each 6 ml of fresh osteogenic medium (*see* **Note 15**).

3. On day 3 after seeding, transfer the constructs to each of the perfusion bioreactors for culture in osteogenic medium for up to 5 weeks. Check the bioreactors for potential leaks, and tighten loose connections. Transfer the seeded bone constructs into culture chambers (1 construct per culture chamber, 6 constructs per round bioreactor presented in Fig. 1) using sterile forceps. Cover the culture chamber with small amount of DPBS to prevent drying, and check the fluid flow through the constructs by attaching a syringe with fresh osteogenic medium into the perfusion loop and pushing the medium toward the bioreactor chamber with constructs (*see* **Note 16**). Fit the constructs into the chambers tightly to allow the

flow through all parallel chambers, and remove any bubbles blocking the flow using forceps and syringes.

4. Aspirate the DPBS and add an appropriate amount of osteogenic medium into the bioreactor medium reservoir (~6 ml per each construct, a total of 40 ml per round bioreactors in Fig. 1).

5. Cover/close the bioreactor chambers, wrap with parafilm, transfer to the incubators, and attach to the peristaltic pump to start perfusion.

6. Adjust the medium flow rate according to the selected study parameters. For instance, a uniform flow rate of 3.6 ml/min corresponding to an interstitial velocity of 0.8 mm/s was used in our studies (14, 15). The flow rate was calibrated and set using a digital, low-flow, multichannel Masterflex peristaltic pump (Cole Palmer). Culture medium in the bioreactors is recirculated and maintained in equilibrium with the atmosphere in the incubator (standard conditions).

7. Monitor the bioreactors daily for any potential leakage or flow obstruction.

8. Along the experimental period, exchange 50 % of the osteogenic medium volume twice weekly with fresh medium, and collect medium aliquots for biochemical analyses (*see* **Note 17**).

9. Harvest perfused bone constructs and statically-cultured control constructs after 3 and 5 weeks of culture (or other selected time points), weigh, cut, and process the samples depending on the selected biochemical and histological analyses (14, 15).

4 Notes

1. During initial expansion of hPSC stocks for storage in liquid nitrogen, we usually omit the antibiotics from the culture medium.

2. We usually test one vial of MEFs from each lot to assess the cell survival after thawing, and the ability to support undifferentiated hPSC growth.

3. Acetone fixative solution is prepared fresh prior to each staining.

4. Faxt blue RR Salt solution is prepared fresh prior to each staining.

5. AgNO$_3$ solution can be stored at room temperature protected from light.

6. Use drills of appropriate strength to avoid tip breakage. We used drills made of Copper 40–50 %, Iron 40–43 %, Chromium 1–3 %, and Tin <1 % (purchased from Sampson Diamond Tool).

7. Differences in derivation efficiency and quality of derived progenitors can be observed when using different hPSC lines. 1 ng/ml of basic fibroblast growth factor can be used to boost MP expansion.

8. Do not let the samples dry. Samples can be stored in PBS at 4–8 °C for later evaluation.

9. Use a white background to reflect light. When exposing to light, keep the samples in H_2O to prevent drying.

10. Control cultures typically present with minimal or absent staining.

11. Make sure no traces of bone marrow are left in the bone matrix before continuing.

12. Polishing assures the scaffolds fit tightly into the bioreactor chambers, forcing the medium flow through the interior of the scaffolds. Slightly less-perfect scaffolds can be used for static culture controls.

13. During initial MP expansion, we usually note the cell growth so we can plan accordingly to expand sufficient numbers of cells until the start of bioreactor experiment. Per each culture/control scaffold, 1.5 million cells are needed.

14. At this high cell density, the cell volume is significant. We usually estimate the volume of cell pellet after centrifugation, and then add an appropriate volume of medium to achieve the correct final seeding density.

15. When planning a bioreactor experiment, an appropriate number of scaffolds and cells is dedicated to various controls.

16. Use medium with phenol-red as an indicator of the fluid flow.

17. Due to evaporation, total volume of the medium in the bioreactor chamber is measured each time by aspirating into a 50 ml pipet. 20 ml of the spent medium is returned into the chamber, and 20 ml of the fresh medium is added.

Acknowledgements

This work was supported by the New York Stem Cell Foundation-Helmsley Investigator Award (to D.M.); the Leona M. and Harry B. Helmsley Charitable Trust; Robin Chemers Neustein; Goldman Sachs Gives, at the recommendation of Alan and Deborah Cohen; New York State Stem Cell Science Shared Facility, Grant C024179; National Institutes of Health Grants DE016525 and EB002520, (to G.V.-N.); and the New York Stem Cell Foundation.

References

1. de Peppo GM, Marolt D (2013) Modulating the biochemical and biophysical culture environment to enhance osteogenic differentiation and maturation of human pluripotent stem cell-derived mesenchymal progenitors. Stem Cell Res Ther 5:106

2. Tandon N, Marolt D, Cimetta E, Vunjak-Novakovic G (2013) Bioreactor engineering of stem cell environments. Biotechnol Adv. 7:1020–1031

3. Marolt D, Knezevic M, Vunjak-Novakovic G (2010) Bone tissue engineering with human stem cells. Stem Cell Res Ther 2:10

4. Tachibana M, Amato P, Sparman M, Gutierrez NM, Tippner-Hedgesm R, Ma H, Kang E, Fulati A, Lee HS, Sritanaudomchai H, Masterson K, Larson J, Eaton D, Sadler-Fredd K, Battaglia D, Lee D, Wu D, Jensen J, Patton P, Gokhale S, Stouffer RL, Wolf D, Mitalipov S (2013) Human embryonic stem cells derived by somatic cell nuclear transfer. Cell 6:1228–1238

5. Takahashi K, Tanabe K, Ohnuki M, Narita M, Ichisaka T, Tomoda K, Yamanaka S (2007) Induction of pluripotent stem cells from adult human fibroblasts by defined factors. Cell 5:861–872

6. Murry CE, Keller G (2008) Differentiation of embryonic stem cells to clinically relevant populations: lessons from embryonic development. Cell 4:661–680

7. Kuznetsov SA, Cherman N, Robey PG (2011) In vivo bone formation by progeny of human embryonic stem cells. Stem Cells Dev 2:269–287

8. Levi B, Hyun JS, Montoro DT, Lo DD, Chan CK, Hu S, Sun N, Lee M, Grova M, Connolly AJ, Wu JC, Gurtner GC, Weissman IL, Wan DC, Longaker MT (2012) In vivo directed differentiation of pluripotent stem cells for skeletal regeneration. Proc Natl Acad Sci U S A 50:20379–20384

9. Fröhlich M, Grayson WL, Wan LQ, Marolt D, Drobnic M, Vunjak-Novakovic G (2008) Tissue engineered bone grafts: biological requirements, tissue culture and clinical relevance. Curr Stem Cell Res Ther 4:254–264

10. Grayson WL, Fröhlich M, Yeager K, Bhumiratana S, Chan ME, Cannizzaro C, Wan LQ, Liu XS, Guo XE, Vunjak-Novakovic G (2010) Engineering anatomically shaped human bone grafts. Proc Natl Acad Sci U S A 8:3299–3304

11. Grayson WL, Bhumiratana S, Cannizzaro C, Chao PH, Lennon DP, Caplan AI, Vunjak-Novakovic G (2008) Effects of initial seeding density and fluid perfusion rate on formation of tissue-engineered bone. Tissue Eng Part A 11:1809–1820

12. Grayson WL, Marolt D, Bhumiratana S, Fröhlich M, Guo XE, Vunjak-Novakovic G (2011) Optimizing the medium perfusion rate in bone tissue engineering bioreactors. Biotechnol Bioeng 5:1159–1170

13. Marcos-Campos I, Marolt D, Petridis P, Bhumiratana S, Schmidt D, Vunjak-Novakovic G (2012) Bone scaffold architecture modulates the development of mineralized bone matrix by human embryonic stem cells. Biomaterials 33:8329–8342

14. de Peppo GM, Marcos-Campos I, Kahler DJ, Alsalman D, Shang L, Vunjak-Novakovic G, Marolt D (2013) Engineering bone tissue substitutes from human induced pluripotent stem cells. Proc Natl Acad Sci U S A 21:8680–8685

15. Marolt D, Campos IM, Bhumiratana S, Koren A, Petridis P, Zhang G, Spitalnik PF, Grayson WL, Vunjak-Novakovic G (2012) Engineering bone tissue from human embryonic stem cells. Proc Natl Acad Sci U S A 22:8705–8709

16. de Peppo GM, Sjovall P, Lennerås M, Strehl R, Hyllner J, Thomsen P, Karlsson C (2010) Osteogenic potential of human mesenchymal stem cells and human embryonic stem cell-derived mesodermal progenitors: a tissue engineering perspective. Tissue Eng Part A 11:3413–3426

17. de Peppo GM, Svensson S, Lennerås M, Synnergren J, Stenberg J, Strehl R, Hyllner J, Thomsen P, Karlsson C (2010) Human embryonic mesodermal progenitors highly resemble human mesenchymal stem cells and display high potential for tissue engineering applications. Tissue Eng Part A 7:2161–2182

Methods in Molecular Biology (2014) 1202: 185–187
DOI 10.1007/7651_2014
© Springer Science+Business Media New York 2014

INDEX

Printed by Printforce, the Netherlands